E. A. Arn

Group Technology

An Integrated Planning
and Implementation Concept for
Small and Medium Batch Production

Springer-Verlag Berlin Heidelberg GmbH 1975

Ernst Alexander Arn, Ing. (grad.), Ph.D.
Staff Manager of Production Engineering and Work Study of Sulzer Brothers Ltd., Winterthur (Switzerland)

With 134 Figures

ISBN 978-3-540-07505-9   ISBN 978-3-642-66262-1 (eBook)
DOI 10.1007/978-3-642-66262-1

Library of Congress Cataloging in Publication Data
Arn, E. A. 1928 - Group technology. Bibliography: p. Includes index. 1. Group technology.
I. Title.
TS155.A685 658.5'33 75-34078

Setting: R. & J. Blank, München

Preface

Group Technology (GT) as a manufacturing concept has gained steady interest within the machine building industry all over the world. Originally it was used more or less only in the so-called parts family manufacturing concept. With growing opportunities for using the computer in the design process, operating planning and layout planning, the potential advantages became more and more obvious.

In order to implement GT successfully and with a view to improving the overall economic situation of a production company, it is necessary to consider all aspects of the complete manufacturing system.

Experience has shown, that in the first stage a general basis has to be formed. This is done by a clear and practical definition of three GT manufacturing systems, the development of a set of classification systems for work pieces, working operations and manufacturing equipment, and in building a data bank from which a data basis for the GT planning process can be evaluated.

A second stage of implementation then considers the particular aspects of GT. These are, firstly, a concept for layout and investment planning based on a representative parts spectrum; secondly, for application of the GT-idea in the design process three similar types of parts are to be developed as a logical supplement to the standard and recurring parts practice; thirdly, a three stage process planning and work measurement system can be developed for the so defined spectrum of similar parts.

Finally consideration is to be given to the aspects of production control for GT and a quantity-quality linked bonus wage system for group work.

The concept put forward in this book is the result of long years of study and of experience in an important Swiss machine building company at the same time considering scientific work carried out mainly by English and German Universities.

I thank especially the Management of Sulzer Bros. Ltd. Winterthur for the opportunity given to me to carry through these investigations in their works, this goes mainly to Dipl. Ing. O. Hegi and Dipl. Ing. K. Vogel.

Furthermore I am indebted, to Professor R.H.Thornley and Senior Lecturer J.H.F. Sawyer of the University of Aston in Birmingham, England, as well as to Dr. H.-P. Wiendahl, University of Aachen, Germany for the many constructive discussions and proposals.

For the assistance in testing the concepts I would like to give my cordial thanks to Messrs. W. Andres, R. Brechbühl, R. Frey and A. Trachsel. Dipl.Ingl. R. Sägesser and Mr. D. Luff were of great help in translating the work into English.

I also wish to thank my dear wife for the great understanding and for the active support at any time.

Hinwil (Switzerland), January 1975   E. A. Arn

Contents

1 Introduction

Group Technology (GT) has been recognized as a method for rationalizing small and medium batch production for some time. One of the first considerations to be made on this subject was put forward by Mitrofanov [1]. With the increasing possibilities of being able to analyse and evaluate the resultant data with the aid of electronic data processing equipment, this method is becoming a practicable solution.

The term 'Group Technology' or 'Parts Family Manufacture' signifies a method which endeavours to analyse and to arrange the parts spectrum and the relevant manufacturing process according to the design and machining similarity so that a basis of groups and families can be established for rationalizing the production processes in the area of small and medium batch sizes. To utilize some of the advantages of rationalization in small and medium batch production, GT makes use of the production engineering techniques that are applied for mass production. Nevertheless, the immediate aspects of mass production, e.g. the reduction of machining times, are not so prominent as the indirect effects on the overall economy such as reducing the throughput time, reduction of work in progress, reliable deliveries to stock, the simplification of production control and the rationalization of design and planning processes.

The effects of GT on the various factors of the corporation are shown in Fig. 1.1, which has been compiled by Thornley [2]. The preliminary measures that have to be adopted to attain these achievements are complex and require a comprehensive concept which permits a gradual but integrated implementation. Such a complete concept must therefore be considered from several aspects, whereby grouping, in respect of the machining requirements and design, must be feasible to begin with.

The following will briefly describe the partial problems of both points of view according to their direct or indirect aims.

In the case of the indirect partial problems on the design side, the reduction of the parts variety through the standardization of the parts having the same or similar shape and functional features is particularly evident. As a result of this, a pre-condition is created for the application of computer aided design. At the same time, these measures also serve to raise the similarity degree in respect of machining requirement characteristics which, in turn, has an effect on the layout of more rational manufacturing systems and on the standardization of process planning.

One of the immediate partial problems is the build-up of data for recurring parts as well as the preparation of pre-printed drawings and a micro-film card index so that the searching and drawing effort expended in the design stage can be reduced. Owing to the lack of qualified draughtsmen and the resultant relatively high time factor involved — in relation to the entire production process — greater importance should be attached to this group of problems.

The indirect partial problems in respect of manufacture include the rationalization of investments and layout planning. The object of this is to increase the reliability of esti-

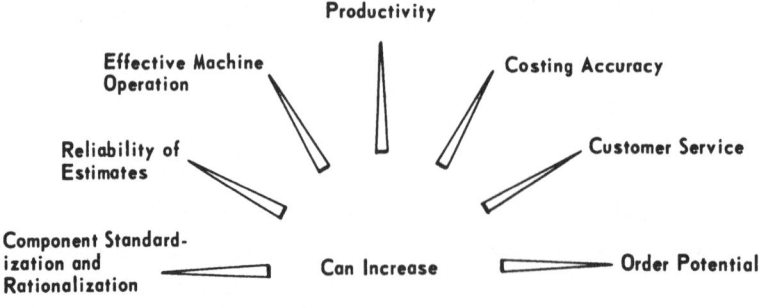

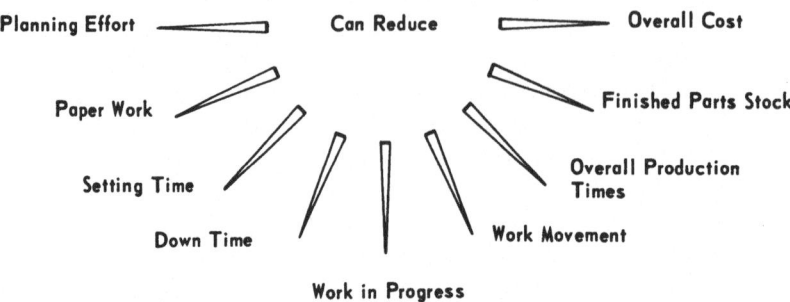

Fig. 1.1. General Achievements of Group Technology

mates and reduce the effort needed to formulate the planning data with the aid of data processing programmes. Apart from the design characteristics the machining similarity characteristics describing the process and the existing production equipment are used as a basis. The immediate problem in respect of manufacturing considerations includes the development of process planning techniques in accordance with GT principles. This partial problem represents a continuation of the layout planning and its purpose is to prepare data for technological and economic decisions concerning process planning for the assigned GT-manufacturing system.

Further partial problems exist in the build-up of technological data for production control so that similar parts may be included in the manufacturing process with a view to increasing productivity by reducing the setting times and the training times for machine operators. At the same time, this data should also provide the production control with information as to the alternative possibilities in the event of machine breakdowns. A further important factor is finally the adaptation of the work measurement and the wage structure to the GT-manufacturing systems in order to ensure the overall economy of this form of rationalization. With these partial problems, the psychological and sociological knowledge of the work forms has also to be considered. These are e.g. the principles of job rotation, job enlargement and job enrichment.

The literature contains a large number of remarkable contributions in respect of these partial problems. Over the last few years, ways have been elaborated at the Technical University of Aachen to solve the design-orientated problems [3]. Different theoretical

2

and application-orientated investigations have been published in respect of the GT-manufacturing systems and various empirical papers have been presented at conferences for Group Technology in Essen (1963, 1967, 1969), Turin (1969) [4], at the University of Aston in Birmingham (1971, 1973), as well as at the CIRP Conference in Sweden (1972) [5], at the University of Manchester [6] and at the Group Technology Center in Blacknest.

In an important Swiss machine building company a number of GT-layouts have been developed and these have been operating for a few years now. The family groupings of components for these layouts were carried out by peripatetic ocular means and it soon became apparent that there were serious limitations to these approaches. It therefore became essential to elaborate the fundamental problems of the individual parts and experience showed that in addition a unified consideration was needed for the complete rationalization of design and manufacture for GT. This work should include the organizational and sociological aspects as well, so that the advantages of GT may be utilized to the full when it is introduced into the company.

The objective of this work therefore is to produce a system-concept whereby the individual components are directed towards the overall objective of GT and coordinated with other rationalization techniques of industrial engineering to form an efficient manufacturing system.

This problem is considered by means of systems engineering, which has already proven its worth as a useful aid for the comprehensive consideration in other scientific fields.

In accordance with these preliminary considerations, the following steps are taken:

— Elaboration of the concept for comprehensive establishment and integration of GT principles in a complex production programme in mechanical engineering.
— Testing these concepts within the framework of selected product examples and manufacturing areas.
— Revising these concepts on the basis of the test results and creating a general procedure.
— Presenting a system-concept for the introduction of GT.

Different product areas formed the basis of the investigation. These areas are characterized by the development and manufacture of high technical quality industrial products with a wide product programme and a diverse manufacturing structure. The products selected for investigations differ in their functions and quality requirements as well as in the batch sizes and the manufacturing techniques.

The concept for planning and implementation of GT is structured according Fig. 1.2:

The external influential factors concerning GT, systems engineering and the new scientific knowledge in the field of production engineering constitute the foundation for the work.

The system concept may be subdivided into two crucial areas. The first provides for the general aspects. The basic forms of GT-manufacturing systems represent the starting position.

A comprehensive classification and coding system for workpieces, operations and equipment is used to create a general planning basis for the development of the various aspects of GT. A technological data bank, based on representative product types, is created for a rational evaluation.

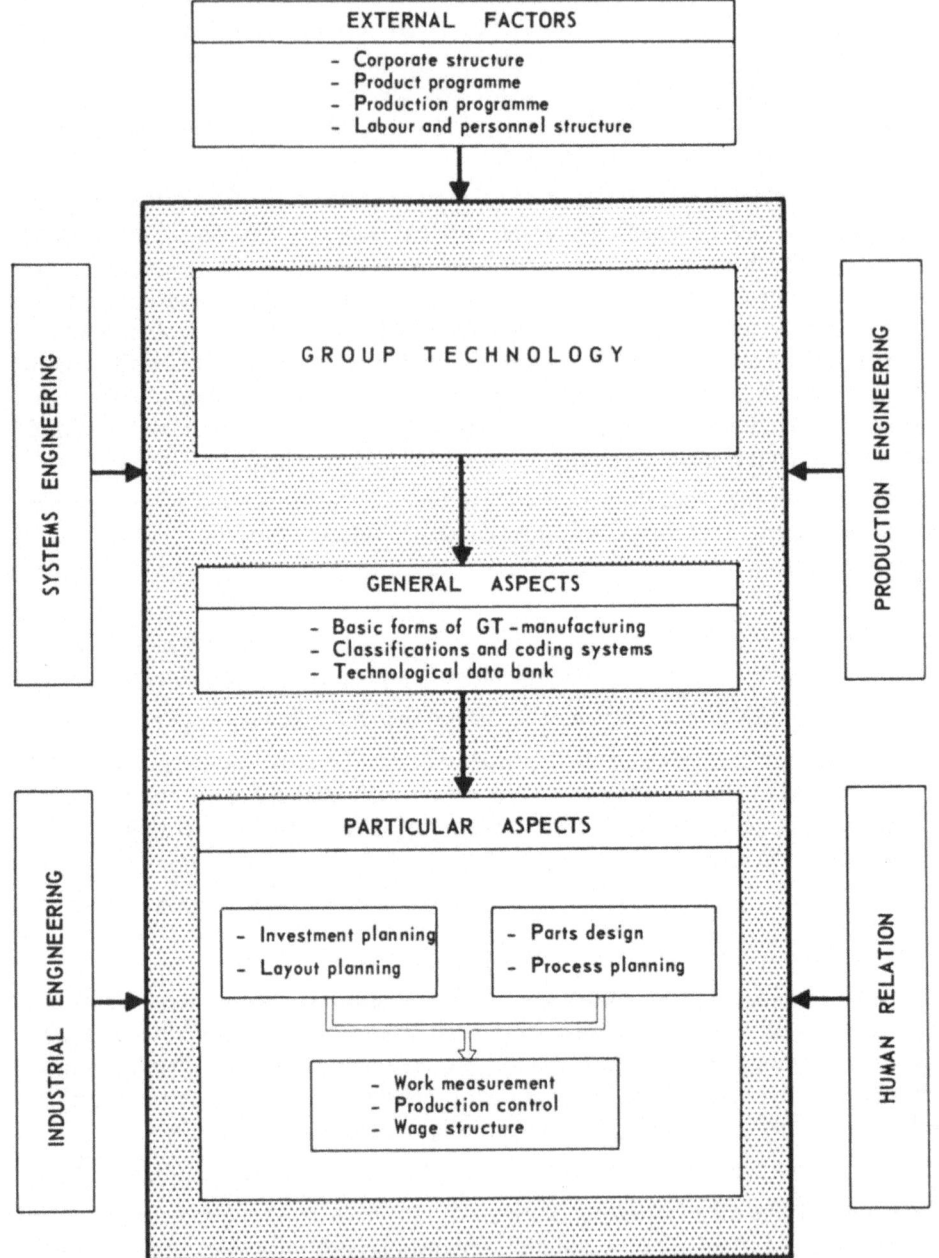

Fig. 1.2. Survey of the Work

In the second crucial area, the particular aspects influenced by GT are dealt with by the way of procedure principles and model examples. Apart from the investment and layout planning, the parts design and process planning, the subsequent problems of work measurement, the production control and wage structure are included because they are also important for the successful introduction of GT.

4

2 The Basic Forms of GT-Manufacturing Systems

From the literary references and practice, there are a number of well known manufacturing processes to which different manufacturing systems can be allocated. Depending on the order batch size (small or large batches), functional machine layout and flow line systems (Fig. 2.1) are already recognized in the industrial manufacturing process.

The basic idea of parts family manufacture originally consisted of grouping parts with similar machining characteristics together to form so-called 'additive batches', and routing them through the functional machine layout with the assistance of the produc tion control.

Functional Machine Layout System

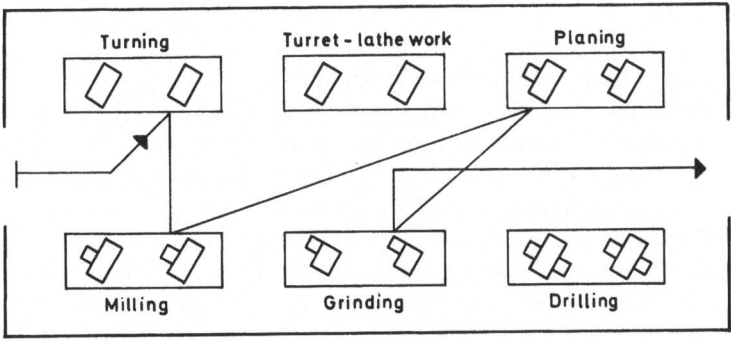

L = Lathe
M = Milling machine
P = Planing machine
D = Drilling machine
G = Grinding machine

Flow Line System

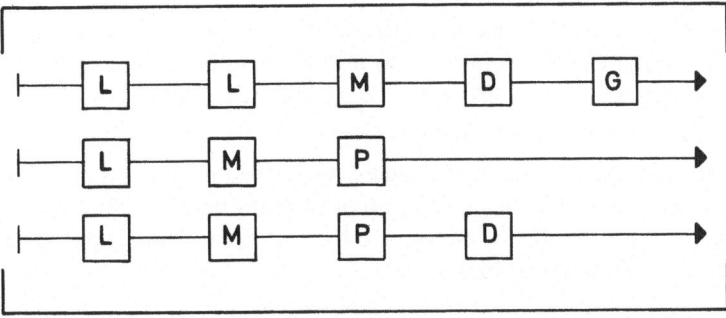

Fig. 2.1. Functional Machine Layout and Flow Line System

In a further development of this idea, parts spectrums with similar machining requirements were set up and executed in corresponding machine groups. With due consideration of the literature [4, 7, 8, 9], GT-manufacturing system can be related to the following basic forms:

- the GT-Centre (single machine system),
- the GT-Cell (group layout system),
- the GT-Flow Line.

The three layout forms lie between the functional machine layout — as the characteristic layout for one-off and small batch production — and the flow line system as the representative of large batch production.

The GT-manufacturing systems differentiate themselves principally through the degree of similarity of the parts in respect to manufacturing. The term 'GT-manufacturing System' includes therefore not only the layout form, but also all the necessary measures, e.g. parts design, process planning, work measurement, production control, wage structure and the psychological and sociological aspects of the work forms, e.g. job enrichment.

2.1 GT-Centre

The GT-centre is developed from the functional layout and consists of a place of work which, from the technical and economic view point, is laid out in such a way that a similar parts spectrum in respect to machining can be executed by a same type of operation, e.g. turning. The application area for the GT-centre provides for a parts spectrum with a similarity in one type of operation and which can be carried out at one individual workplace or on one machine. It therefore constitutes the first and lowest degree of rationalization within the framework of the GT-manufacturing systems. Fig. 2.2 provides a schematic representation of the GT-centre and also shows its significant aspects, machines, parts spectrum and the respective similarity characteristics within the framework of the functional layout.

This basic form has become again increasingly important due to the technical progress of flexible transport systems, e.g. the principle of inductively-controlled tractors, whereby the direction of motion is controlled by means of an a.c. guide conductor laid under the ground. The tractors are controlled from central stations (distribution centres) and in contact with the assigned work places in the manufacturing shop. Some German firms are already using this system.

The first planning objective of the GT-centre lies in the layout of the production equipment and the determination of the most favourable workplace with the assistance of a similarity analysis of the parts manufacturing. This is made in relation to economic investigations on the basis of relevant cost elements.

The next objective is the rationalization of the paperwork needed for process planning data. This problem is associated with the building up of planning data and the gradual implementation toward computer-aided process planning, e.g. in the form of operation sequence plans, tool lists and presetting tables for tools.

A further partial objective is the reduction of setting and training times through similar machining assignments on the respective equipment in accordance with sequence plan-

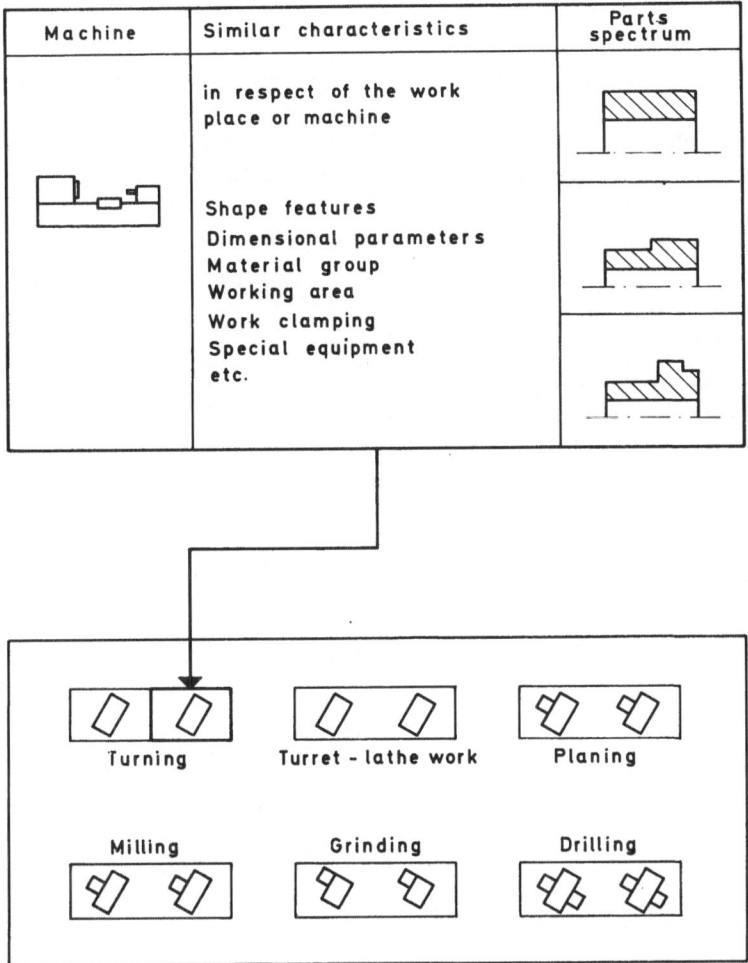

Fig. 2.2. GT-Centre

ning principles. With the aid of GT-centres, it is now possible to create for production control a better sequence of information and data sheets in the form of sequence families via the technological similarities of the work pieces. In particular cases, it is also possible to extend the same to cover the preparation of tools and other auxiliary devices.

The numerically-controlled machine tools and particularly the machining centre occupy a special place within the framework of a GT-centre. Apart from the optimum conformity between machines and workpieces, the main emphasis in this case lies in the setting up of NC-programming families with a view of rationalizing the programme build-up. This is also the case when universal and computer assisted programming systems are employed. The rationalization effect in this field of application lies in the grouping together of different parts with similar programming requirements and this is expressed in the development of limited programme instructions and the specialization of the programmer.

2.2 GT-Cell

The basic idea of the GT-cell is to split the manufacturing area into machine groups in which all the machining operations required for the manufacture of a certain parts spectrum can be accomplished. This basic form of GT-layout allows a flexible operation sequence and constitutes a second or medium degree of rationalization. Fig. 2.3 illustrates the principle of a GT-cell.

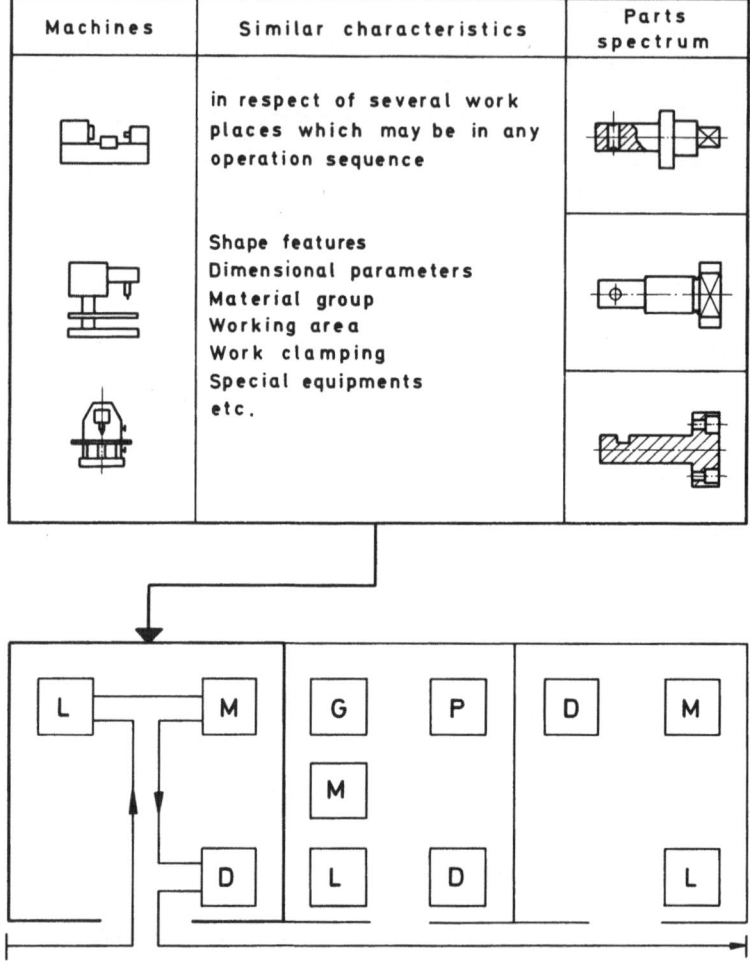

Fig. 2.3. GT-Cell

In addition to the rationalization objectives of the GT-centre, the reduction of through-put times and a more simplified form of production control also come to the forefront here. The layout form also gains in importance from the technical management view-point, because it constitutes a self-contained but limited area of responsibility and thus allows quality assurance to be effected in a more simplified manner.

This method also has a positive effect on the sociological aspects, for it provides a good basis for the formation of work groups with common objectives.

2.3 GT-Flow Line

In the GT-flow line, the places of work for the associated parts spectrum are laid down in a layout according to a fixed operation sequence (Fig. 2.4). This case thus represents the third or highest degree of rationalization for GT-manufacturing systems. It enables the transport between individual places of work with the respective handling equipment, depending on the parts characteristics and the peculiarities of the work shop, to be organized in a rational manner. If various machining times occur at the single places of work during the throughput, corresponding buffer areas and alternative machines must be provided. This also applies in cases where the optimum utilization of an expensive key machine is called for.

The GT-flow line comes very close to the rationalization advantages of mass production. Apart from its application possibilities in the case of specific product flow lines, e.g. crankshaft production lines, this basic form can also be of economic importance for the grouping together of more simple parts such as flange components, which present

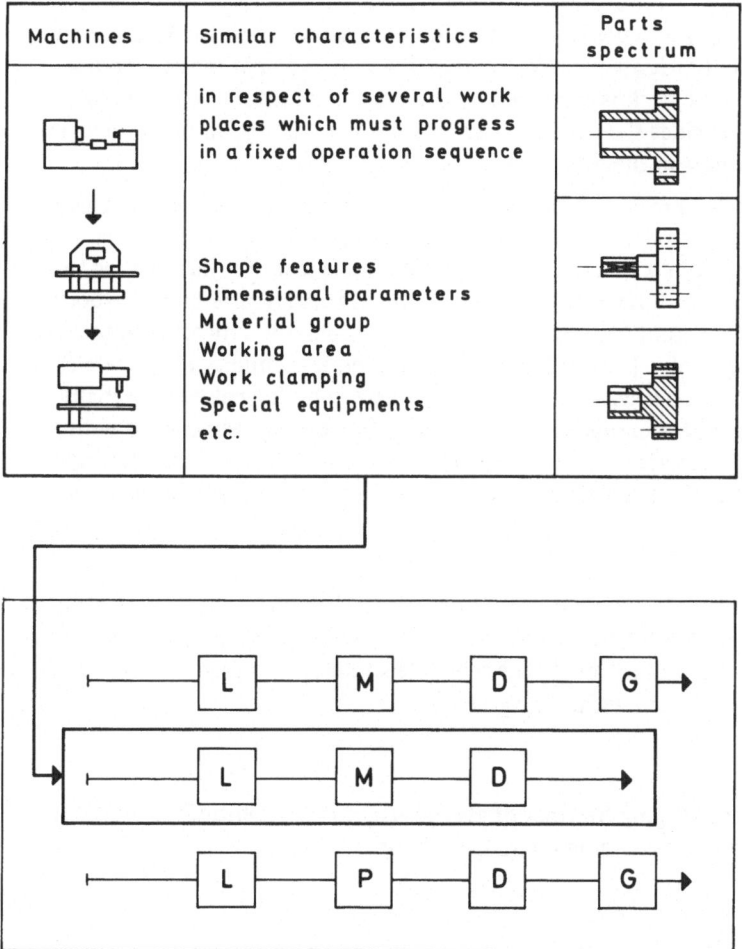

Fig. 2.4. GT-Flow Line

no specific product characteristics. With this basic form, as well as these of the GT-cell, the work places are no longer loaded up individually. Depending on the application case, a periodic work programme for example is sufficient for the flow line. If the flow line contains a differentiated parts spectrum, the similarity characteristics of the individual parts are listed in the work programme. In this way production control is provided with technological information, as an aid for work loading in the most favourable operation sequence.

2.4 Psychological Working Aspects of GT-Manufacturing Systems

These considerations are based on Taylor's principle of work division and the replaceability and interchangeability of the employee. At present, this is generally practised in the manufacture of large batches in order to attain maximum productivity.

Opposed to this are the basic ideas of the human relation movement which place the human relationships in the foreground.

The integration of the two principles results in the behaviour science movement, which is becoming increasingly accepted and aims at the humanization of working life and acceptable productivity. Consequently, the required improvement in the working situation at the place of work is expressed by the increased latitude in respect of human action. This enables the worker to activate his mental capabilities with a view to increasing his interest in the work and his own satisfaction.

The increased latitude in respect of human action is dependent on the existing personnel structure, the educational level as well as the social-psychological aspects and the prevailing social structure. In matters relating to the basic ideas for humanizing working life, signs of a certain fundamental change in the organization of human labour in industry have been apparent for some time. This is particularly noticeable in the automobile industrie, e.g. Volvo [10] and electrical industry, e.g. Philips [11], where the principle of work division is abandoned. Here, the assembly line has been partially abolished and replaced by an organizational form with working teams. Even with purely manual work, the idea is to activate the mental capabilities and the feeling of responsibility for the job. The following will briefly explain the possible work forms:

Job Rotation

This implies a procedure for a working group in which the employee A takes over the work of the employee B who, in turn, assumes the activities of employee C in accordance with a specified or self-selected time schedule so that a complete change round of all places of work is effected within the group.

Job Enlargement

This appertains to the ranging together of several structurally identical or similar work elements, with a view to extending the scope of work.

Job Enrichment

Job enrichment entails the grouping together of structurally different work elements, e.g. planning, manufacturing and inspection acitivities, to form a larger action unit.

10

Autonomous Work Group

In this case, the so-called autonomous work group is allocated a certain task. The group then makes its own arrangements for accomplishing this task. The clarification, distribution, execution and control activities are assigned to the group and not individual persons. Such an experiment has been tried in Norway [12].

GT constitutes a good technological basis for the application of the above mentioned form of work. It is founded on the endeavour to group the manufacturing process of similar part groups together in a unified manner, which is comparable to job enlargement. The extent of the work involved constitutes an important criteria with group work.

From the technological point of view, a claim is often made for larger manufacturing cells so that better flexibility and loading of the machines is attained. From the psychological working aspect, however, smaller groups are recommended because with the increasing size, the reciprocal influence on the group objective and the solidarity is reduced.

The linking of the above mentioned forms of work with the layout forms of the GT-manufacturing systems depends a great deal on the personnel structure. In the case of structures with a good general and professional standard of education, it must be inclined more and more towards job enlargement, enrichment and autonomous work group. Such tendencies show that an association between GT-manufacturing systems and work forms cannot be established because they are influenced considerably by the respective structure of the personnel.

The following provides a relevant recommendation which compares the layout forms of GT-manufacturing systems with the described forms of work with a view to optimizing productivity without considering the existing personnel structure.

The GT-centre is particularly advantageous when associated with job rotation, where similar work is grouped together and the training time is kept within economic limits. Job enrichment can be accomplished by introducing self-inspection for qualified personnel. Within the GT-cell itself, all the forms of work can be employed with the advantage that the task area is limited in such a way that the members of the group also have the feeling of belonging to a team.

The GT-flow line, which is very near to the layout forms for large batch production, presents the same psychological working disadvantages. With the aid of job rotation or job enrichment by means of self-inspection and the setting up of the machines by the operators, the latitude of human action can be extended.

To sum up, it can be stated that in comparison with the functional layout, the GT-manufacturing systems present a better starting point for the step by step introduction of the recommended forms of work.

3 Systems Engineering in the Production Process

The term 'Systems Engineering', is understood to be a methodology which enables known and new scientific methods and techniques to be represented in a schematic form so that they can be employed in gaining knowledge of complex facts, and in developing and realizing new systems. Systems engineering enables complex problems to be broken down into individual partial problems in a logical manner and to investigate same without losing the interrelationship to the overall problem [13].

When applied to GT, it is advantageous to subdivide the complete process within a production company into problem-orientated system areas which represent a limited area of activities. The formulated area of systems shown in Fig. 3.1 represent the first assignment stage. The specific tasks and activities of the production process are

— Design,
— Process Planning,
— Manufacturing and Assembly.

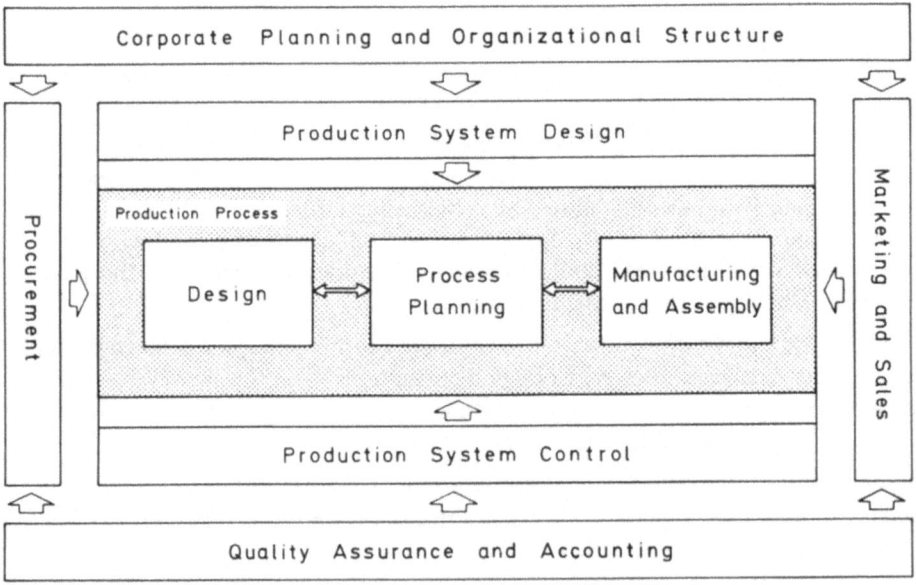

Fig. 3.1. Systems Area of Corporate Level

They form the basis for the realization level of the corporate objective. The direct system areas, which are subdivided into a field of Production System Design and Production System Control are grouped around this realization level of production processes.

The indirect system areas

— Corporate Planning,
— Procurement,
— Marketing and Sales,
— Quality Assurance and Accounting

cover the external influential factors of the corporate process.

The following describes the individual system areas in brief:

— Corporate planning and organizational structure deals with the long-term and mid-term planning of corporate aim and the resultant structurizing of the corporation to reach the planned objective.
— Production system design has the task of attuning the methods and processes which contribute to the rational realization of the technical and economic overall objective in design, process planning and manufacturing.
— Production system control is to optimize the throughput times of the orders.
— Procurement deals with the optimum material arrangements in respect of times and costs.
— Marketing and sales covers the comprehensive marketing and sales promotion systems.
— The quality assurance and accounting systems are auxiliary aids for the control function of the technical and economic corporate function of the technical and economic corporate objectives.

3.1 Area of Production System Design

As GT lies more or less within the influence sphere of production system design, this system area will be examined much more closely. Various methods and techniques are now available which enable the respective problems to be solved both systematically and rationally. From the systems engineering viewpoint, however, the comprehensive consideration does not exist when applying these methods individually. This is especially noticeable within the framework of technically high-quality products where, as a consequence of the ever increasing requirements, work has to be specialized and divided accordingly. Furthermore, the various partial problems become even more dependent on each other because of technical progress. Overall considerations are therefore called for to enable individual techniques to be applied in a comprehensive manner. An analysis of the various rationalization problems of production system design showed that they can be subdivided into three part areas in the first stage in such a way that they form a basis for the techniques and working aids and also correspond to the three essential components of the production process, namely product, equipment and the human element.

The objective of product design is to raise or assure the technical value of product not only in the systemization of the design process, but also in the build-up of work aids for the rational elaboration of drawings and planning data. Thus, the term product design describes the comprehensive application of methods and work aids during the design process of a product to the release of the basic data for process planning. The objective of manufacturing design is to systemize the planning and preparation processes for the manufacture and assembly and to build up the planning data for the rational realization of the different partial assignments within the framework of the concept.

The aim of work study and wage structure is to optimize the working conditions at the work place and to increase output in accordance with scientific methods and processes.

On the basis of these objectives, a structural concept was developed for the production system design in the field of mechanical engineering, and the fundamental methods and working aids were coordinated accordingly (Fig. 3.2). As a result of the described objective analysis, the crucial points shown in the plan represent the first stage. They constitute the basic form for further classification stages and will be described in greater detail in the following. The build-up represents a system hierarchy which is comparable to the functional structure of a product.

Even if certain methods are suitable for dissimilar tasks, it is still advantageous from the practical point of view to associate the crucial points of the area of activities with a certain rationalization objective.

GT is a special case because it has a wide objective spectrum, as is shown by Thornley in Fig. 1.1. The problem will be explained in more detail in the explanations concerning the individual part areas and the connection with GT will be demonstrated.

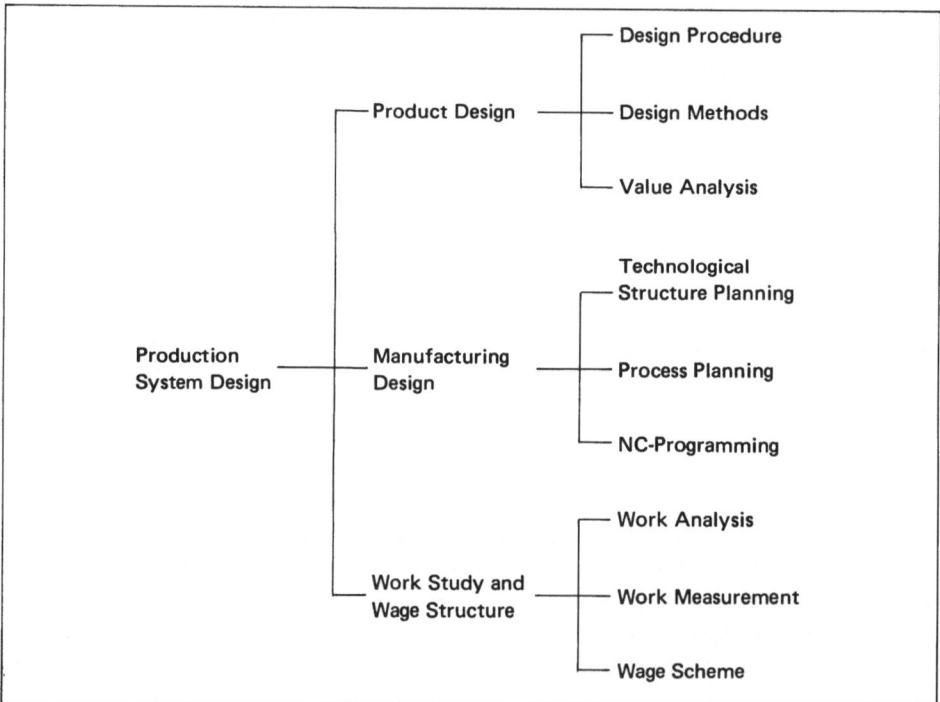

Fig. 3.2. Arrangement of Production System Design

3.1.1 Product Design

The systemization of product design and the increased employment of methods and working aids become more and more important when it is realized that the total costs of the product are influenced to a decisive degree during this stage, starting from the

14

Cost Structure   Cost Influential Factors

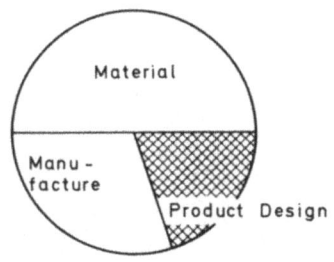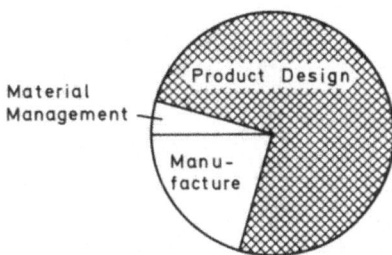

Fig. 3.3. Cost Structure and Cost Influential Factors

assumption that it must be differentiated between the cost structure and cost influence. If only the costs caused by the various branches are considered then the conclusion will be reached that the product design does not constitute any special crucial point of rationalization. However, if it is assumed that at this stage the selection of the solution and the resultant decisions made in regard to material, dimensions and accuracy have a very great influence on manufacture, then it will be quite clear that the product design is a decisive economic factor. These two aspects are schematically compared with each other in Fig. 3.3.

A further crucial point lies in the rationalization of the recurring design activities to shorten the throughput time. Various investigations with individual and similar designs in the field of mechanical engineering have shown that manufacture and assembly are only responsible for 30% - 50% of the total throughput time, whereas the preparation activities such as sales, design, process planning and procurement account for 50% - 70% of the time. Within the actual design stage, a statistical analysis of eleven firms showed the following time shares for the various activities [14]:

16 % for Conceiving of Design,
15 % for Outlining,
55 % for Elaborating,
14 % for diverse Activities.

This is illustrated in Fig. 3.4. On the basis of the development' order, the conceiving include the activities required for clarifying the tasks, the abstraction and formation of the task into sub-functions, finding solution principles for the subfunctions and co-ordinating these principles so that they comply with the overall function. Further design variations in the form of sketches based on the combinations of the selected principles and the technical-economic evalutions are used as a preparation of decision for determining the best solution. Outlining includes the development of alterna-tive scale designs based on this best solution and the technical-economic evalua-tion of these and the selection and determination of the most favourable outline which is used as the fundamental for the elaboration phase. Elaboration is concerned with optimizing the established design zones of the selected outline, the preparation of the drawings and releasing to the process planning department.

The corporate planning which covers the clarification of the long term product develop-ment and the respective activities constitutes the connection with the product design. The activities include, for example, the investigations concerning the integration of GT in the production process for the purpose of increasing economic efficiency of

15

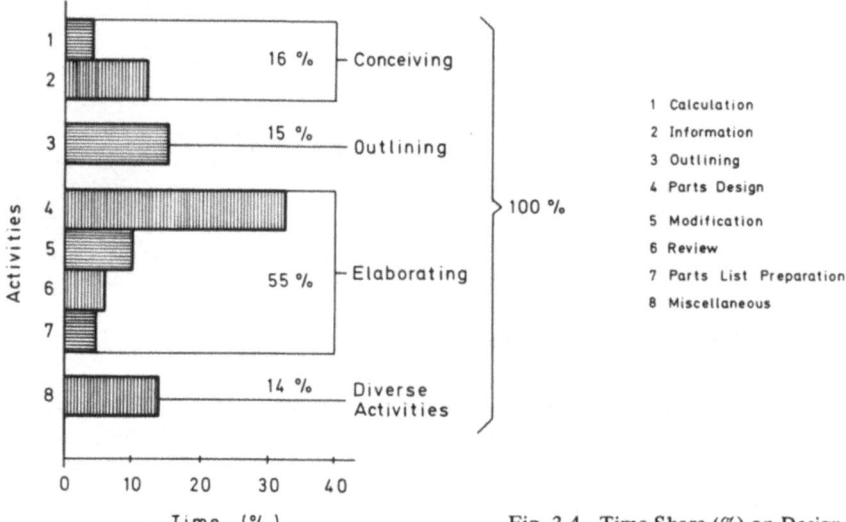

Fig. 3.4. Time Share (%) on Design Activities

1 Calculation
2 Information
3 Outlining
4 Parts Design
5 Modification
6 Review
7 Parts List Preparation
8 Miscellaneous

products with small batches. On the basis of this situation, the following crucial points can be deduced in the case of product design:

— Development of a general procedure for the product design with a view to coordinating the methods and working aids in the best possible manner.
— Improvement of the working aids in order to obtain quicker information in respect to design solutions by using a framework of similarity data and data of assembly groups for GT.
— Further extension of the technical economic design with decision data on the basis of relative costs.
— Increased employment of static and dynamic computer programmes for determining components and assembly groups.
— Drawing of certain single parts with the aid of graphical display unit on the basis of similarity data.

Fig. 3.5 shows the design stage — conceiving, outlining and elaborating of the product design and the interconnection to the other two rationalization crucial points — manufacturing design and work study. The first GT influence takes place in the phase of conceiving by recommendations concerning characteristic limiting values in respect of machining dimension and requirements. In this way we establish the basis for applying GT-manufacturing systems with a higher degree of rationalization right at the conceiving and outlining stage of a product. A further influence occurs in the elaborating stage, which includes the part design of the product. This can be effected by means of GT design guide lines, recurring parts, standard parts, similarity types and the resultant pre-printed drawings as the preliminary stage for computer-aided design. The similarity types are especially dealt with in chapter 6. Close connection between GT and value analysis should be aimed for at this level.

By this way, an optimization will be obtained between manufacture and function. The individual steps of the product design, together with methods and working aids to be recommended, are described in the framework concept under the term 'Design Proce-

16

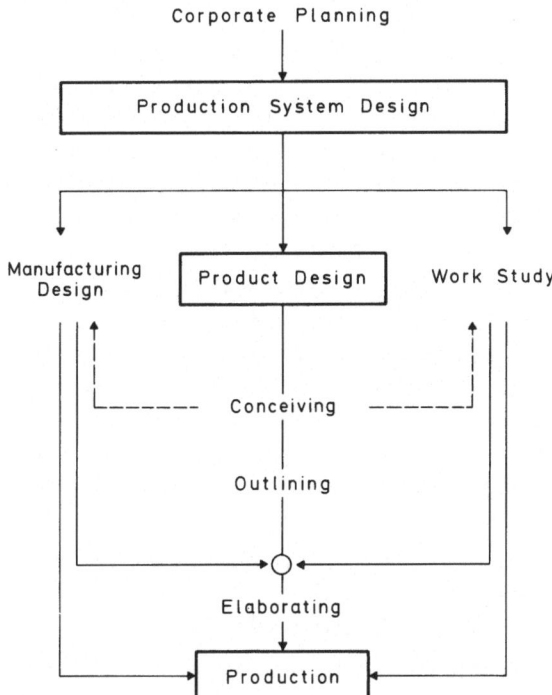

Fig. 3.5. Structure of Product Design

dure' (Fig. 3.2) [15]. The selected methods for mechanical engineering, e.g. brain-storming, morphological methods, technical-economic evaluation methods and value analysis, are further crucial rationalization points, which are presented as limited subsystems in the system arrangement plan.

To a similar effect, the working aids of GT have to be incorporated in this system area. By doing this, the designer can be influenced in an objective-orientated manner during the design process of a product. Fig. 3.6 illustrates such an assignment plan.

Product Design		Work Aids for Group Technology		
		Guide Lines	Recurring Parts Standard Parts	Similarity Types
Design Stage	Conceiving	X		
	Outlining	X	X	X
	Elaborating	X	X	X

Fig. 3.6. Table of the Working Aids within the Product Design

3.1.2 Manufacturing Design

With increasing mechanization and automation, the various design problems in the manufacturing stage become more and more interrelated. Apart from the specific tasks like process planning and NC-techniques, the structural problems such as the manufacturing solutions for the mid- and longterm production programme, within the framework of the corporate planning, the creation of a unified basis for investment and layout planning as well as the technical associations for the product design and work study, all play a decisive part within the framework of manufacturing design.

The assigment area of the manufacturing design is therefore subdivided into levels for the purpose of rational and step-by-step solution of the various planning activities. The GT planning basis for the solution of general tasks constitutes a main level and is referred to in the following as 'Technological Structure Planning'. The main level of the specific production tasks includes the particular problems of process planning and NC-techniques. Systems in the form of procedure and working aids, which duly account for the associations with the other planning levels, have to be created for the different planning levels.

Fig. 3.7 shows the relation between the mentioned levels of manufacturing design and also the connection with other system complexes. Within the framework of GT, the general aims of technological structure planning can be presented as follows:

— Investigation of the technological main criteria — workpieces, operations and equipment — for the crucial points of rationalization.
— The setting up of a comparable planning basis with the aid of a unified classification and coding system to solve tasks that are different, but nevertheless have relationships with one another in design, process planning and manufacture. Whereas the technological structure planning system covers the overall planning problems and should be particularly effective in the case of general planning tasks, e.g. investment and layout planning, the process planning systems deal with the objective-orientated partial problems. The partial general process planning methods, for example, the presetting of tools, the cutting data, etc., cover besides the conventional process planning both GT and NC-techniques.
— The partial area similarity planning methods represents a further factor in the case of small and medium batch production. Apart from the shape, machining characteristics and further technical influential factors, the batch size, the annual parts consumption and other criteria have a great importance on the selection of the best possible form of machining.

The GT-manufacturing system therefore provides the starting point. Here, the objectives are:

— Forming of machining families by the collection of parts with similar operations and operation sequences.
— Coordination of the parts and the optimum machining equipment with the assistance of data concerning machine groups.
— Rationalization of the places of work through standardization of group equipment.
— Increasing the degree of utilization by grouping together work-pieces which are similar from the point of view of machining requirements.

NC-machine tools require more and more integrated system solutions, which influence other branches of process planning. In this particular case, for example, the sub-systems

18

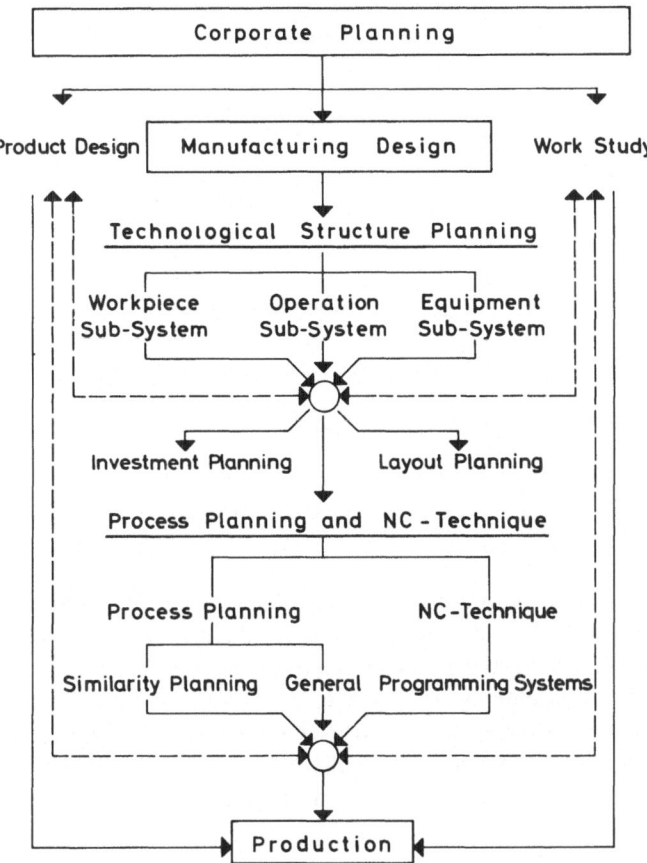

Fig. 3.7. Structure of Manufacturing Design

presetting of tools and cutting data are becoming more and more important for conventional manufacturing, NC-techniques and GT.

The partial area programming techniques has the task of providing data in order to control the NC-machine and to give further instructions to assist processing, e.g. tool-lists.

The partial area programming technique can be generally divided into two main groups and sub-divided yet again into two further groups [16] (Fig. 3.8), namely

Without languages:
— Programming without a programming language and without computer, but with programmable desk computers,
— Programming without a programming language, but with mini-computers;

With languages:
— Programming with specific programming languages or symbols and mainly with mini or medium size computers,
— Programming with a universal programming language and mainly large sized computers.

19

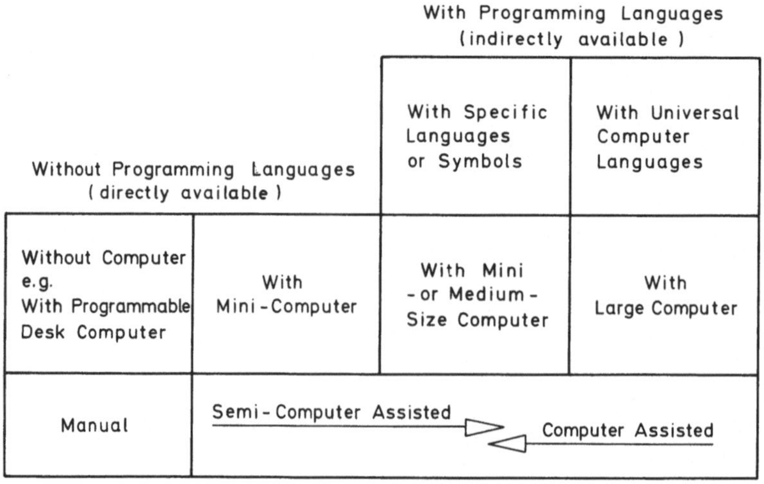

	Without Programming Languages (directly available)	With Programming Languages (indirectly available)	
		With Specific Languages or Symbols	With Universal Computer Languages
Without Computer e.g. With Programmable Desk Computer	With Mini-Computer	With Mini- or Medium- Size Computer	With Large Computer
Manual	Semi-Computer Assisted ⟶		⟵ Computer Assisted

Fig. 3.8. General Arrangement of Programming Systems for NC-Machine Tools

In future, the universal programming language will probably become more predominant, because the number of languages for the individual problems would have to be increased and there would be too many different programming languages to master. Consequently, sub-systems for special cases of machining will have to be deduced from a general programming language. A planning basis may be formed therefore by grouping together the workpieces with similar programming requirements in so-called NC-programming families. A further development may be seen in the graphic display and screen programming system.

3.1.3. Work Study and Wage Structure

This partial area includes systems which are principally based on scientific working knowledge. They comprise the various techniques of work analysis, work measurement and wage structure. In the development of the systems and research methods employed in this area, the technical statistical processes become increasingly important. Apart from the pure systems engineering factors, the multilayered environmental influences of the relationships between human being and the technique have also to be considered.

In connection with GT, the optimum assignment of the methods for workmeasurement connected with the GT process planning levels are particularly in the forefront. If output is to be statistically assured then the measurement of quality and quantity performance also constitutes a further problem area within the framework of GT-manufacturing systems.

3.2 Systems Engineering Structure

The ever-increasing degree of mechanization and automation within industry necessitates a strengthening of the planning and designing activities. Furthermore, an increased number of marginal conditions have to be considered nowadays when solving individual functional problems. The time factor plays a decisive role. A reduction in the time between the product idea and its realization is required with a simultaneous increase in the complexity of the products. In order to comply with the additional requirements in due time, this development resulted firstly in a further work division. It soon showed, however, that this objective-orientated specialization for the autonomous fulfilment of individual tasks has certain limits. For example so many safety factors were built into the various partial solutions that the economics were eventually no longer in accordance with the overall optimum. On the basis of this, integrated methods and work aids were called for in planning and designing.

The terms planning and designing can be defined as follows:

Planning, therefore means to pre-determine all the steps needed to reach a certain aspired objective and to establish the ways and means with which the objective can be reached under the best possible conditions in respect of time, cost, work technique, planning data and manpower [17]. Designing represents the creative part of the accomplishment of a task and its objective is to establish the necessary conditions for the optimal realization. Planning and designing thus constitute a constant reciprocal action. Design is not only limited to the product development, but it refers to the entire production process with the functions planning, conceiving, outlining, elaboration, preparation and operation.

By using methods to solve production tasks, the area of activities of the engineer has become larger. In the past, his main task was to understand the physical and technical principles and to use them for the solution of the different problems. Nowadays, in solving such problems, he has to determine their direct and indirect area and to optimize their interaction. The solving of such problems calls for knowledge of various fields and consequently leads to increased teamwork.

Generally speaking, systems engineering as a suitable tool in the field of industrial engineering has been developed gradually. It began with the work study as a staff function. The objective of this activity was to rationalize the work in the factory with aid of scientific and technical methods. As a result of the increase in comprehensive consideration of the production process the investigation work has also spread to preparation areas of design and process planning.

An optimum design logic stands in the forefront of systems engineering whereby the relationship of the various part tasks have to be duly considered. The objectives of the specialist and system functions which constitute the partial tasks are shown in Fig. 3.9.

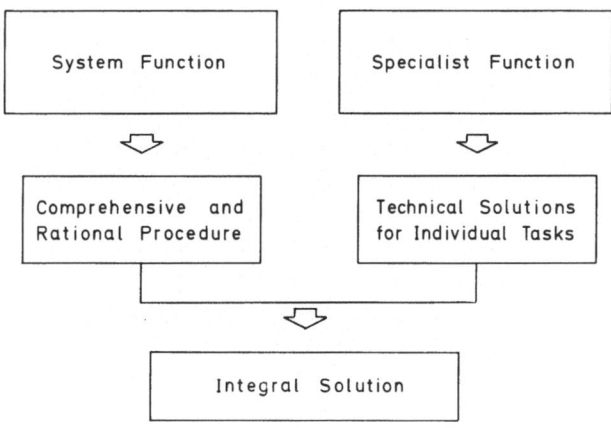

Fig. 3.9. Specialist and
System Function Objectives

21

The specialist gives priority to the functional realization in due course of his individual objective, whereas the systems engineer is more interested in the optimization of the overall objective. This difference is particularly important in the realization of complex systems as here the systems must be employed beyond the boundaries of the functional areas and the development of the solution methods must therefore be both neutral and centrally organized. The engineer must therefore master the systems engineering methods if he is to fulfil this task. This requirement determines the procedure and organization between the specialized staff and systems engineering. In the development of systems within the production system design area, the marginal conditions of the other task areas and the knowledge of various scientific fields has to be integrated in such a way that the influential factors of the other activities are duly considered. This is the only way if the tasks are to be accomplished autonomously and according to plan. With production system design, the knowledge of design and manufacturing techniques and the principles of work study, technology, management organization and plant economics as well as applied mathematics play a decisive role (Fig. 3.10).

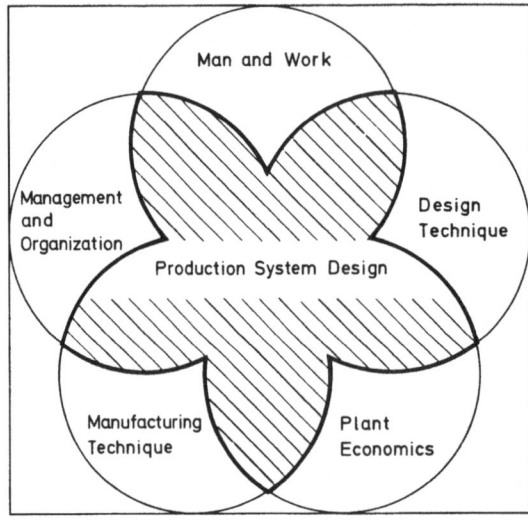

Fig. 3.10. Influential Factors for Production System Design

3.3 Systems Engineering Procedure

The systems engineer can also consider the techniques and working aids as products and in their development employ procedure techniques and solution methods similar to those used for the product design. The starting point is the systems engineering conception of production system design which constitutes a limited problem area and a link in the comprehensive system structure. In practice, it is customary to disconnect knowingly a part of the entire area so as to allow partial results to be obtained much quicker. Once these partial results have been obtained, they are more or less fed back into the complete system. If, with this procedure, adjustments are required later in view of the further buildup stages, then the chances of success are even greater. In other words, we should try to limit the sub-system within a system in such a way that in accordance with the given financial and personnel means a set system task is solved in stages.

Systems for complex tasks can be used nowadays for a period of about five to ten years. Owing to the rapidly and continually changing structures within the corporations, the systems have to be constantly adapted during this period. Further and new developments occur much quicker now and they should, therefore be regarded as investments and placed on the same level as the production equipment and personal investments.

Apart from the system development, the form of the introduction into the corporate process and the training of future users of the methods and working aids also play a decisive role in the rationalization success and the innovation period of a system. Psychologically, this factor must already be considered in the build-up stage and should also be integrated in the procedure of systems engineering. Because of the inter-disciplinary character of the task, the system design calls for organized and planned co-operation between the specialist staff and the systems engineer, as it is the case with value analysis.

On the basis of this consideration and due to the experience gained with similar projects, the procedure is sub-divided into major stages. The different stages are then split into further steps (Fig. 3.11). This general scheme serves as a foundation for the project organization and the procedure technique when solving complex system tasks.

In the 'preliminary study', it is a question of recognizing the crucial points of rationalization by means of fundamental studies and defining the problem according to a present state analysis. In the case of new developments, a model study is required if it is desired to have a better starting point. This preliminary stage should particularly aid the subsequent build-up and shorten the preparatory work so as to increase the period in which the system may be used and also simplify the transition between theory and application.

The system concept and the relative working methods are developed and tested during the 'system build-up stage'. The system build-up can be compared to the design phase of a product. In the conception, the solution is determined according to a prevously defined procedure and requirements plan. It is during this period that the co-operation between systems engineering and the specialist staff begins.

The 'system training' is characterized by the following three stages which are also determining factors for different information grades:

— Information (general),
— Introduction (technical staff),
— Instruction (specialists).

The objective of the first stage is to familiarize every employee concerned with the system with the fundamentals of the system build-up.

The second stage informs the leading technical staff of the fundamentals and objectives of the system and deals particularly with the application possibilities so as to create pre-conditions for its introduction.

The third stage is confined to the handling of the systems and also includes the transfer of same to the technical staff departments.

The introduction also signifies the start of 'system maintenance'. This infers consultation and information for the technical staff and the further development of methods and working aids.

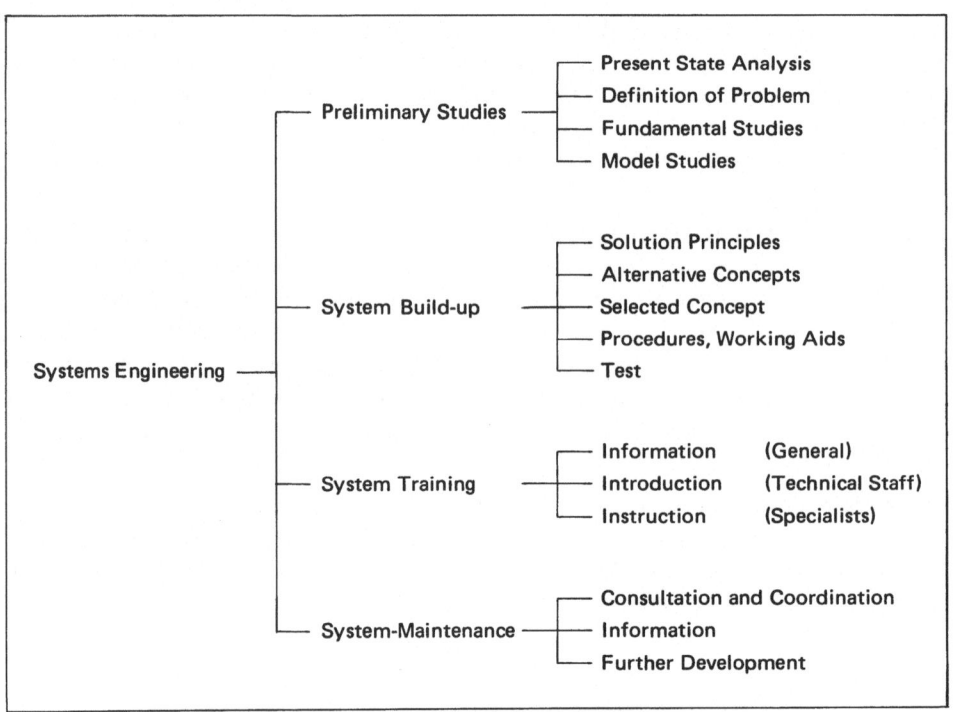

Fig. 3.11. Systems Engineering Procedure

3.4 Rules and Working Aids of Systems Engineering [18, 19, 20]

A system represents the entirety of individual elements that are related to one another. The elements and their association constitute the structure of a system. Systems engineering deals with the procedure needed to develope the system. The following established fundamentals should be observed during the development:

— Systems engineering considers a system as a whole, the parts of which although diversified, are fully integrated with each other and carry out specialized functions.
— Every system has objectives whose relative importance should be considered in every particular case. The elements should be designed and selected in such a way that the best possible overall results are attained in respect of the objectives.
— The main objectives of a system arise from the requirements and necessities of another superimposed and comprehensive system.

Systems engineering is thereby assisted by system theory, operations research and programming techniques.

Systems theory is based on the fundamentals and rules of the structure, the behaviour and the influential parameter of tested systems. Further fundamentals are mathematical methods for optimization of the activity procedures with the assistance of operations research. Operations research means the application of mathematical methods for the problems arising from the mode of operation of systems, e.g. the interplay of human beings, machines and material. In the majority of cases, this refers to the optimization methods with which, if necessary with the aid of a computer, the decision data can be elaborated for the optimum solution. According to the size, importance and kind of problem, mathematical methods such as

— Correlation calculations,
— Linear programming,
— Queueing theory,
— Simulation techniques

may be used.

In the case of simulation, a process is simulated with a model in such a way that the results obtained conform as accurately as possible with the real results. With the assistance of a computer and on the basis of a model test, it is then possible to estimate the previous effect of various influential factors [21].

A further component is the problem-orientated 'programming technique' which, on the basis of build-up logic and the established solution enables us to translate the problem into a computer language.

Even though the concept of systems is to be made separately from the computer technique and is primarily aimed at the solution of the problem, the use of computers in the case of systems with a large quantity of data has to be considered, as it is the case, for example, with GT. Fig. 3.12 shows the combined effect of the influential elements of systems engineering as a technique for rational and integral problem solution, used in the present work.

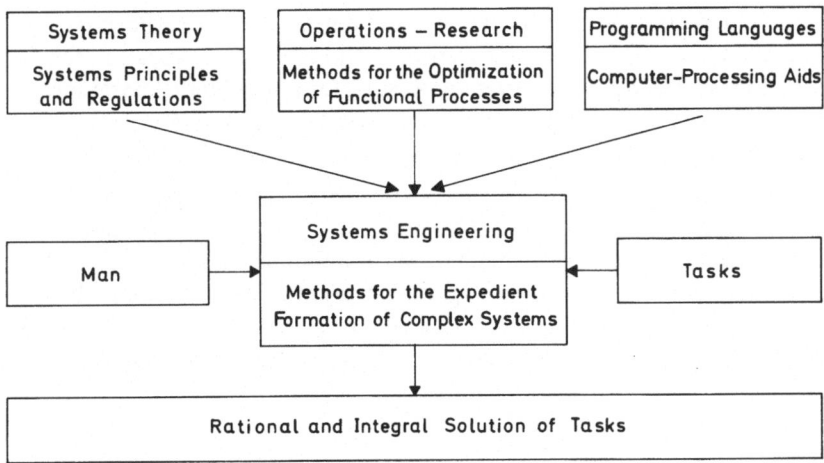

Fig. 3.12. Elements of Systems Engineering

25

4 Classification and Coding

The starting point for effective rationalization of the technological production process is the systematic collection of the relevant technical and economic planning data. Various classification and coding systems have been developed for the purpose of creating a technological planning basis in one-off and small batch production.

If we analyse the functional areas of activities in planning and designing the production process, it is found that a great number of the individual problems have a common planning basis.

From both the technical as well as the time and cost standpoint, the increasing complexity of the planning tasks needs a corresponding systematic if the planning studies are to be effected rationally, unforeseen circumstances to be eliminated and optimum planning and decision data to be drawn up. In view of the large quantities of the data involved, the system should be built up in such a way that various problems can be solved in a rational manner by aid of the computer.

The selection of the task range, the depth of the required conclusiveness and the effort needed to build up and to maintain the system are decisive evalution criteria.

Two fundamental directions are possible in the development of the classification systems. The first consists of building up a respective and specific classification for the individual partial problems, e.g. for a certain process or for a very limited parts spectrum. It can be compared with the specific programming languages used for numerically controlled machine tools.

In the case of the other direction, the objective is to create a planning basis for the various areas of activities in the first stage. This basis, which covers the workpieces, the operations and the equipment, can then be used as a foundation stone for building up further classification systems for specific partial problems. From the comprehensive consideration standpoint, the second direction should be given priority. In the case of the general planning basis, the technological structure analysis of production is more

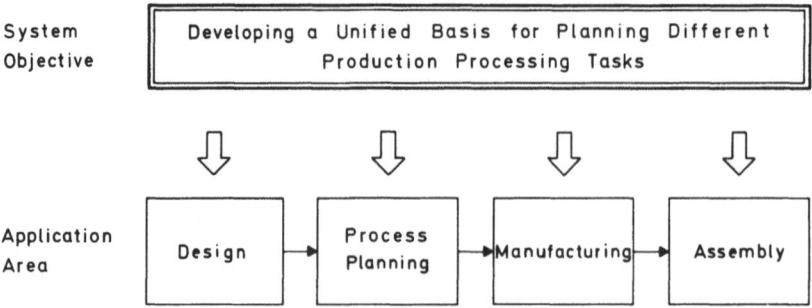

Fig. 4.1. System Objective of Technological Structure Planning

important than the detail problems of the organization and planning of the individual parts. In the following, the creation of this general thesis will be referred to as 'technological structure planning' within the framework of the overall consideration.

The following crucial points stand in the foreground when determining the structure planning objectives (Fig. 4.1):

— Determination of the similar drawings for the build-up of recurring parts and similarity types on the basis of the design families.
— Indication of the crucial point for objective-orientated investment planning for the build-up of GT-manufacturing systems.
— Support for the layout planning by means of reference data.
— Creation of a technological basis for workshop capacity comparison.
— Determination of the similarity structure of parts machining as a basis for the formation of machining families in order to rationalize process planning and equipment utilization.

4.1 Characteristics and Planning Data

In laying down the characteristics and planning data, we restrict ourself to the essential main criteria required for a general technological description of production (Fig. 4.2):

— Similarity criteria of the shape and machining surfaces of the workpieces.
— Material classes with the same technical manufacturing conditions.
— Similarity criteria of the individual operations for the parts machining on the process level.
— Similarity criteria of the application area for equipment on the process level.
— Dimensional parameter depending on the various areas of activities.

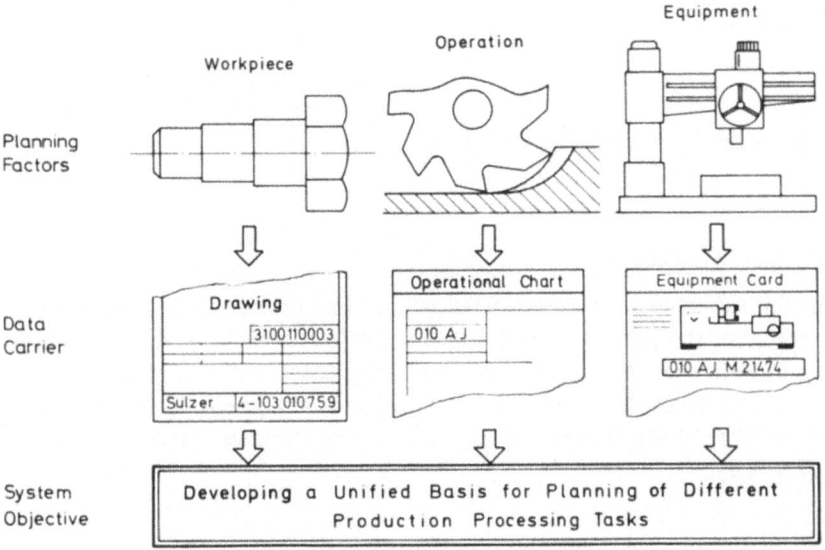

Fig. 4.2. Planning Factors of Technological Structure Planning

4.2 System Structure

On the basis of the described crucial points of rationalization and the resultant main criteria — workpiece shape, manufacturing operations and equipment — the system structure can be subdivided into three separate sub-systems:

— Workpiece (parts structure),
— Operation (operation structure),
— Equipment (equipment structure).

Apart from the classification codes for the description of the workpiece, manufacturing operations and equipment, the main emphasis — as regards system development — lies in linking these three main influential factors to a unified planning basis.

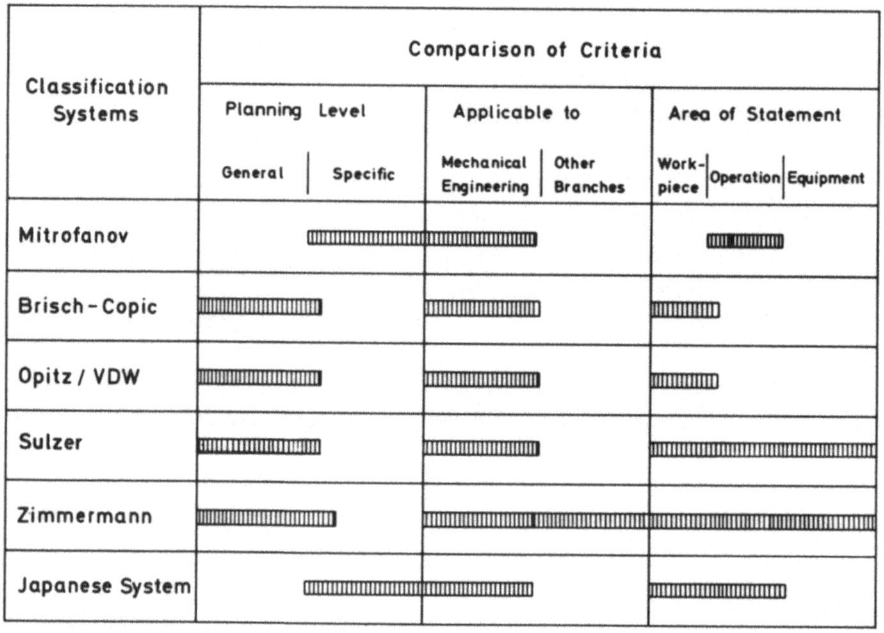

Classification Systems	Comparison of Criteria							
	Planning Level		Applicable to		Area of Statement			
	General	Specific	Mechanical Engineering	Other Branches	Work-piece	Operation	Equipment	
Mitrofanov	▨▨▨▨▨▨▨▨▨▨▨▨				▨▨ ▨▨▨▨			
Brisch – Copic	▨▨▨▨▨▨		▨▨▨▨▨▨		▨▨▨			
Opitz / VDW	▨▨▨▨▨▨		▨▨▨▨▨▨		▨▨▨			
Sulzer	▨▨▨▨▨		▨▨▨▨▨▨		▨▨▨▨▨▨▨▨▨▨▨▨			
Zimmermann	▨▨▨▨▨▨		▨▨▨▨▨▨▨ ▨▨▨▨		▨▨▨▨▨▨▨▨▨▨▨▨▨▨▨			
Japanese System	▨▨▨▨▨▨▨▨▨▨				▨▨▨▨▨▨▨			

Fig. 4.3. Comparison of Different Classification Systems

4.2.1 Sub-System 'Workpiece'

The basis for this sub-system is the classification code for the workpiece description, which also represents the starting point for the complete system. Several known classification systems from the literary references are briefly characterized and then the selected solution will be explained [1, 22, 23, 24, 25, 26, 27]. The following evaluation criteria — in the sense of the determined comprehensive objectives — serve as comparison factors (Fig. 4.3).

— Planning level: general/specific,
— Applicable to: mechanical engineering/other branches,
— Area of Statement: workpiece/operation/equipment.

Further evaluation criteria comprise the simple handling of the system, the classification effort and the computer processing, which are of major importance for practical application in industry. Investigations have shown that some of the systems are very accurate from the theoretical standpoint, but not for their practical applications as they are difficult to incorporate in the existing organization or uneconomic because of the extensive coding effort required. The crucial point, with most of the classification systems investigated, lies in the workpiece description. In the majority of cases, it was assumed that there was a relationship between the shape of the workpiece and the machining. Even if good correlation is obtained with simple workpieces, the degree of determination falls with the increasing workpiece complexity. From the technological planning viewpoint, it should also be noted that two different levels are involved. The workpiece description lies on the part level, whereas the manufacturing process takes place at the operation level. The link up between these two planning levels must be assured in a unified system. In this book, this requirement is met through the aid of the interlinked sub-system workpiece, operation and equipment.

The development of the workpiece description is based on the VDW-classification system devised by the Institute headed by Prof. Opitz [23]. This code was modified in order to maintain a stricter differentiation between the shape and machining description in view of the design and manufacturing technologically-orientated rationalization problems.

The ten-digit workpiece code is used as a working aid to determine the workpiece structure:

The code covers the following planning characteristics:
— Geometric shape,
— Part class (branch between design and manufacture),
— Machining characteristics,
— Dimensional classes.

Fig. 4.4 shows the workpiece code structure. The first three digits are shape-orientated and mainly used for the design stage. The second part of the code chiefly covers the machining characteristics of the workpiece description, because it has been shown that the part classes are very different in consideration of the following machining characteristics and depend on whether cutting, non-cutting or assembly operations are principally involved. The third part contains the dimensional classes. The workpiece is classified according to its two largest dimensions which are determined in accordance with the machining characteristics: rotary components are given the code-digits according to L & D (length and diameter) and non-rotary components the ones according to A & B (longest edges). In this concept, particular attention is paid to the dimensional classes. As a result of frequency of the workpiece spectrum and the equipment working areas, a basic grid consisting of 26 classes each, in accordance with the basic series 'R5' of prefered numbers, has been chosen for the two most important dimensions. To facilitate possiblities of greater variation for the diverse problem statements, various concentration stages are provided in the individual sub-systems. Whereas a coarse concentration stage of 10 classes is provided on the workpiece level, the operation level has 26 — like the basic grid.

Apart form workpieces with cutting and abrasive operations, the structure of the classification system also allows workpiece drawings appertaining to forging, sheet metal and cast components to be coded. As the characteristics are particularly associated with

29

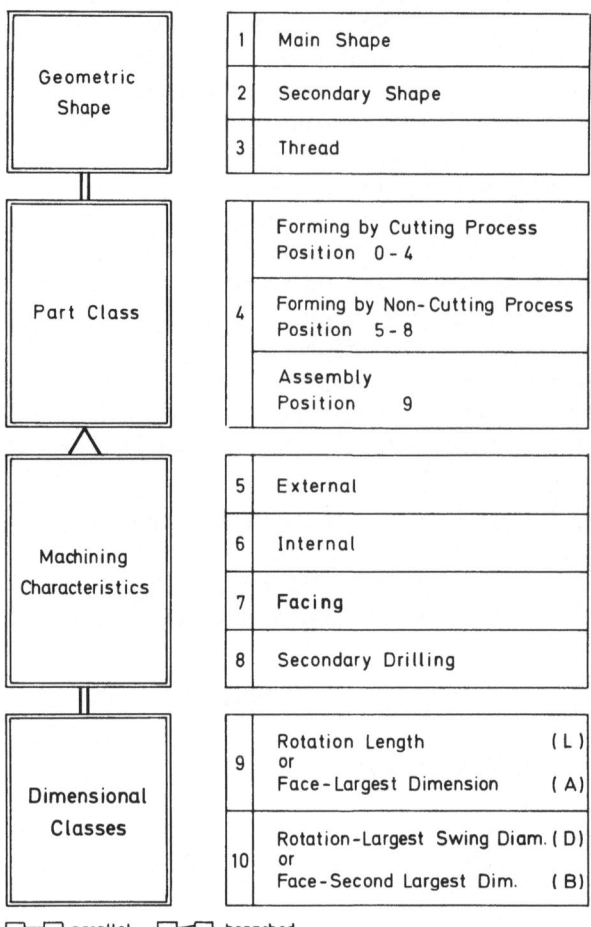

Geometric Shape	1	Main Shape
	2	Secondary Shape
	3	Thread

Part Class	4	Forming by Cutting Process Position 0 - 4
		Forming by Non-Cutting Process Position 5 - 8
		Assembly Position 9

Machining Characteristics	5	External
	6	Internal
	7	Facing
	8	Secondary Drilling

| Dimensional Classes | 9 | Rotation Length (L) or Face-Largest Dimension (A) |
| | 10 | Rotation-Largest Swing Diam. (D) or Face-Second Largest Dim. (B) |

□━□ parallel   □◁□ branched

Fig. 4.4. Structure of the Workpiece Code

the cutting process, the degree of conclusiveness drops off slightly in the case of the non-cutting processes. Ten possible classification positions are provided for each digit. The description and use of such a code is carried out with the aid of special work guidelines and coding examples.

The test phase results showed that it is advantageous to carry out the coding operation centrally as the risk of coding errors is then reduced appreciably. The coding system was tested on more than 10000 workpieces and has now reached the introductory phase.

Fig. 4.5 shows a relevant coding example. The code number is noted in the drawing. For that reason the single part drawing system must be applied. Once the workpiece code number has been entered in the drawing and on the data sheet, it is taken over directly by the computer. In this way, it is possible to carry out an evaluation in accordance with the various grouping standpoints with the existing programmes. The subsystem 'Workpiece' is shown in Fig. 4.6 Apart from the workpiece code, the most important components are the drawing, as the data carrier, the data sheet for the

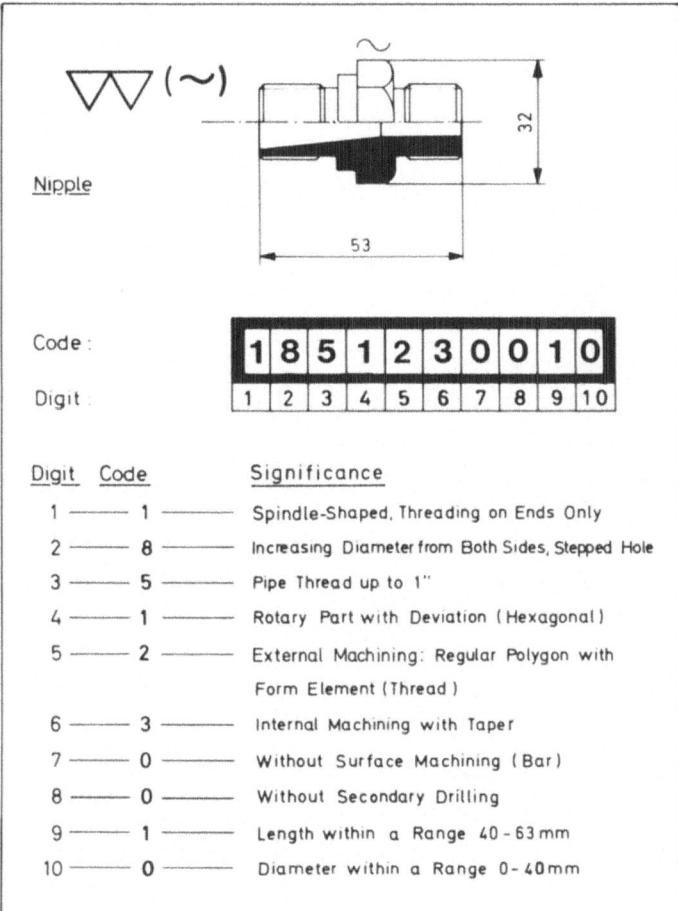

Fig. 4.5. Application of the Workpiece Code

computer input and the evaluation programme for the various problems of rationalization.

Fig. 4.7 represents an evalution of the parts structure from investigated product groups, consisting of diesel engines, compressors and gas turbines. By supplementing with further characteristic analyses, good comparisons are able to be obtained in respect of the parts structure of different products and, on a long-term basis, eventual structural changes within a certain product can be followed up.

4.2.2 Sub-System 'Operation'

The five-digit classification code (Fig. 4.8) is used as the basis for the operational structure. The coding system describes the operations in the operational chart. With the aid of evaluation programmes, the operational code allows a three-stage planning basis to be created independent of equipment. This conception is sub-divided into the following stages:

Coarse — Machining Class,
Medium — Machining Type,
Fine — Machining Process.

31

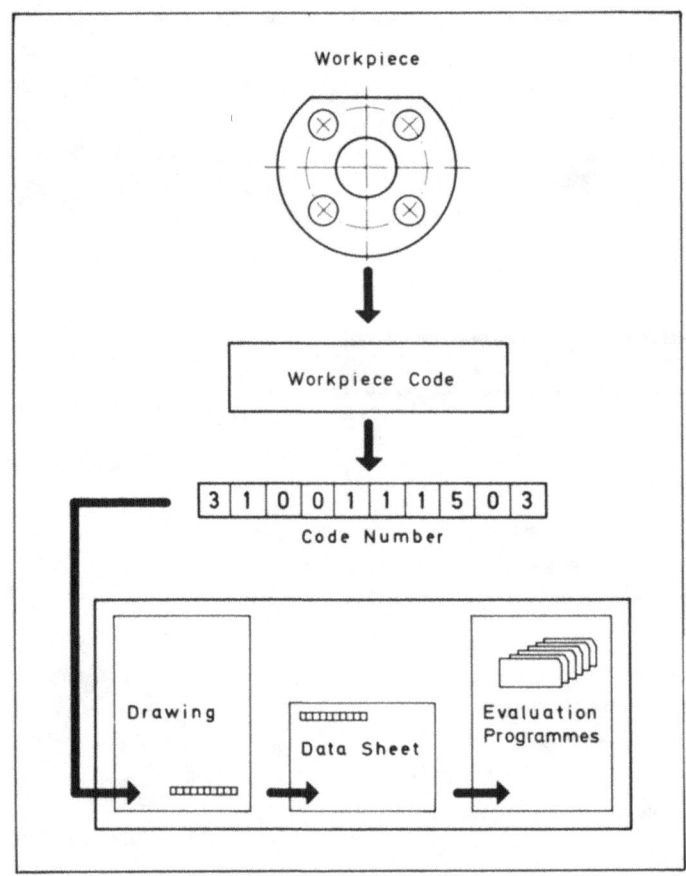

Fig. 4.6.
Sub-System 'Workpiece'

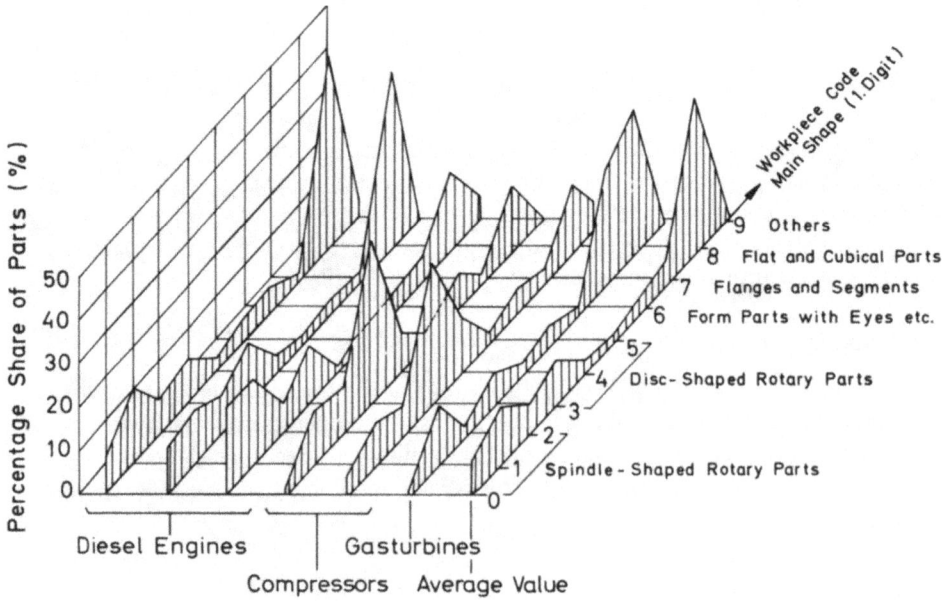

Fig. 4.7. Parts Structure of Different Products

In addition, the two main dimensions of the workpiece — from the machining stand-point, e.g. rotation or non-rotation — are coded for every machining operation. Contrary to the 10-digit workpiece code the 26-digit dimensional grid is used for the operational code. Both dimensional codes are based on the basic series R5 of prefered numbers. The 26 classes meet more or less the requirements of the planning basis for the various rationalization tasks in investment, layout and process planning. In this way, the dimensional classes, which constitute a very important influential factor in the overall system, are more accurately atuned to the individual planning tasks in the manufacturing area, and the degree of change in the operation sequence is also duly considered. In developing the code, particular emphasis was laid on the simple handling so that the codings could be integrated into process planning.

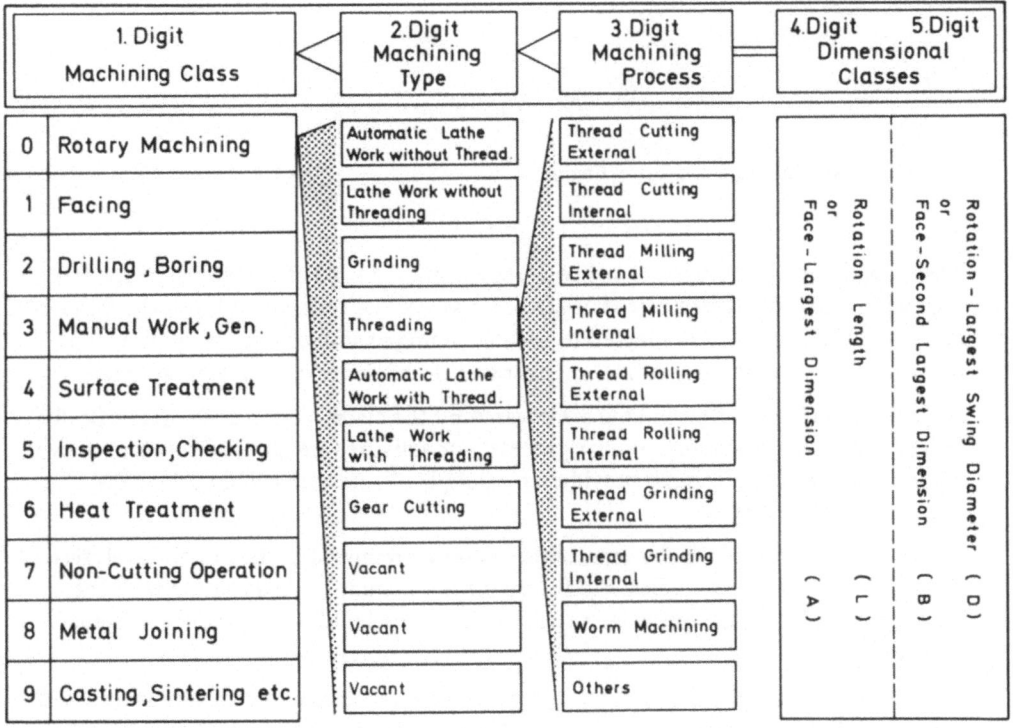

Fig. 4.8. Formation of the Operational Code

The first digit of the operational code covers 10 positions, which describe the cutting and non-cutting processes as well as the surface and heat treatment, checking, inspection and also bench work.

These considerations apply particularly to long-term planning tasks, where the overall aspects of production are in the forefront. Fig. 4.9 represents the coarse stage of the machining structure of a diesel engine. Apart from the parts structure (Fig. 4.7), it gives the planning staff sections their first overall impression, e.g. during preliminary studies of rationalization projects on a large scale.

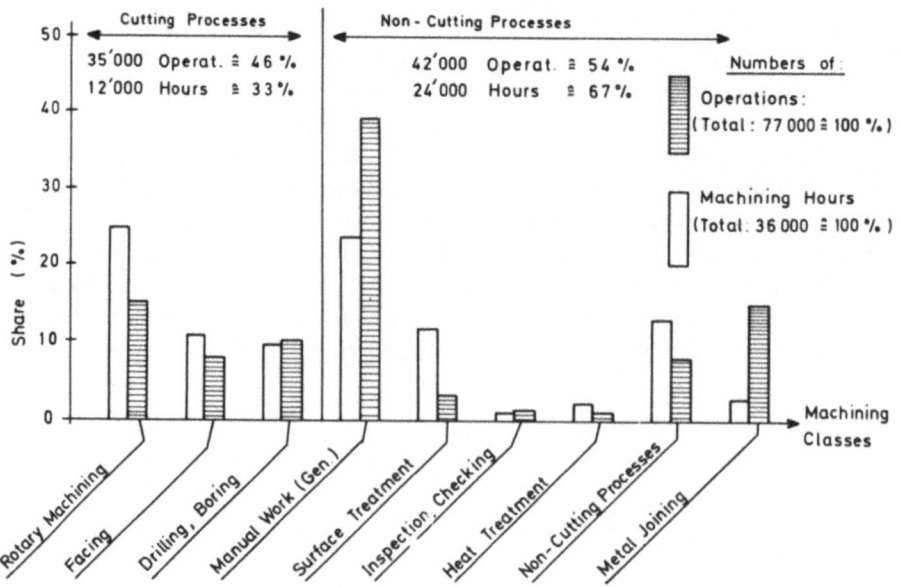

Fig. 4.9. Coarse Structure of a Diesel Engine

The second digit of the operational code represents the type of machining of the first digit and constitutes a refinement of same. However, the refinement is not the same in every machining class. It is adapted to the machining profile of the corporation. The main areas are structured in a finer manner and the adjacent ones in a broader one.

At present, 54 types of machining are covered by the classification code, theoretically 100 possibilities are feasible. Similar processes or activities are grouped together under one machining type. Fig. 4.10 provides an example of rotary machining, whereas Fig. 4.11 shows facing.

The classification level already represents a good working aid in the present analysis of the state for the project studies, e.g. for the layout planning in association with the workpiece coding. It also provides a good basis for the build-up of product reference numbers.

Fig. 4.12 is an example of a product reference table concerning the already mentioned diesel engine. It constitutes a refinement of the course machining structure shown in Fig. 4.9.

The third digit of the operational code describes the individual machining process or in the case of manual work — the activities. It mainly corresponds with the working operations of the operational chart. In particular cases — especially in the pronounced one-off production — the presently used machining process arrangement will have to be planned in a much more detailed manner than it has been the case till now. This requirement, however, also meets the necessity for building up an effective form of process planning and production control.

Fig. 4.13 shows a section of the third stage of the rotary machining area as well as the structure for dimensional classes. The conclusiveness of the machining process already represents a coarse description of the individual operation in the operational chart.

34

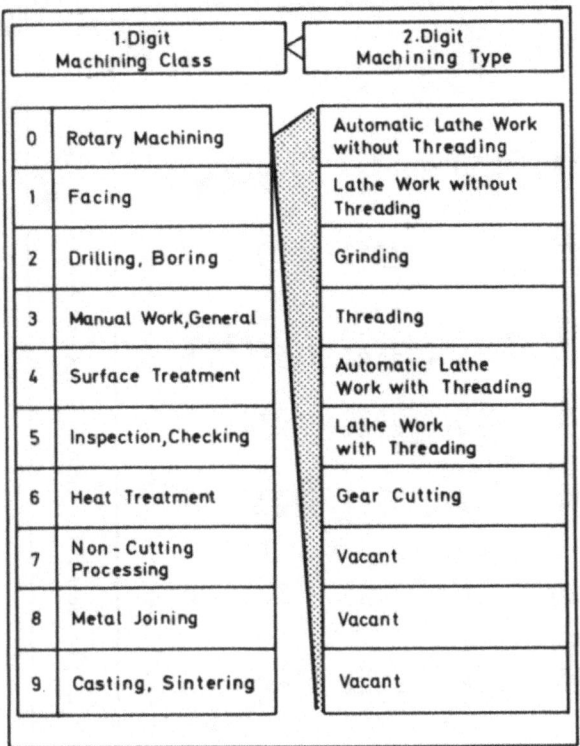

1.Digit Machining Class		2.Digit Machining Type
0	Rotary Machining	Automatic Lathe Work without Threading
1	Facing	Lathe Work without Threading
2	Drilling, Boring	Grinding
3	Manual Work, General	Threading
4	Surface Treatment	Automatic Lathe Work with Threading
5	Inspection, Checking	Lathe Work with Threading
6	Heat Treatment	Gear Cutting
7	Non-Cutting Processing	Vacant
8	Metal Joining	Vacant
9	Casting, Sintering	Vacant

Fig. 4.10. Operational Code – 1st and 2nd Digit – Machining Class and Machining Type (Rotary Machining Area)

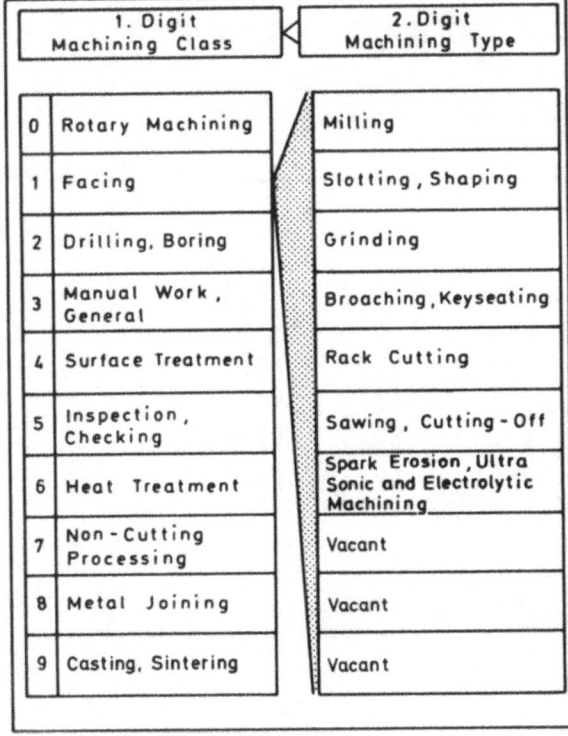

1. Digit Machining Class		2. Digit Machining Type
0	Rotary Machining	Milling
1	Facing	Slotting, Shaping
2	Drilling, Boring	Grinding
3	Manual Work, General	Broaching, Keyseating
4	Surface Treatment	Rack Cutting
5	Inspection, Checking	Sawing, Cutting-Off
6	Heat Treatment	Spark Erosion, Ultra Sonic and Electrolytic Machining
7	Non-Cutting Processing	Vacant
8	Metal Joining	Vacant
9	Casting, Sintering	Vacant

Fig. 4.11. Operational Code – 1st and 2nd Digit – Machining Class and Machining Type (Facing Area)

Dimensional Classes (mm) →	100 / 100	100 / 400	400 / 100	400 / 400	400 / 1000	1000 / 400	1000 / 1000	1000 / 1600	1600 / 1000	>1600 / >1600	TOTAL
	0	1	2	3	4	5	6	7	8	9	
0 0 Rotary Mach. Aut. Lathe Work without Thr.	220.2	224.9	89.8	92.4	0.0	5.4	0.0	0.0	0.0	0.0	632.7
0 1 Rotary Mach. Lathe Work without Thr.	69.1	303.3	184.3	312.3	852.1	164.5	149.8	193.0	144.7	653.3	3 026.4
0 2 Rotary Mach. Grinding	64.4	25.3	98.9	73.4	50.9	43.3	14.5	62.5	0.0	126.5	559.7
0 3 Rotary Mach. Thread Cutting	20.9	0.5	24.8	347.3	4.7	12.0	0.0	13.2	0.0	0.0	423.4
0 4 Rotary Mach. Aut. Lathe Work without Thr.	143.3	1.7	61.0	1.6	4.6	0.4	0.0	0.0	0.0	0.0	212.6
0 5 Rotary Mach. Lathe Work with Thr.	33.0	17.5	36.2	67.8	14.2	103.1	0.0	34.1	0.0	282.7	588.6
Total Hours	560.0	573.2	495.2	894.8	926.5	328.7	164.3	302.8	144.7	1 062.7	5 152.9
1 0 Non-Rotary Mach. Milling	81.7	0.0	111.4	262.8	8.0	120.1	206.4	130.3	1.0	651.7	1 573.4
1 1 Non-Rotary Mach. Planing	0.6	0.0	13.8	56.8	0.0	213.3	137.6	149.2	10.4	143.0	724.7
1 2 Non-Rotary Surface Grinding	3.4	0.0	8.2	93.7	0.0	14.8	26.0	5.1	4.5	22.8	178.5
1 3 Non-Rotary Broaching	4.6	0.0	1.2	30.5	0.0	0.0	0.0	0.2	0.0	0.0	36.5
1 4 Non-Rotary Rack Cutting	32.9	1.7	10.6	85.9	7.8	5.2	77.1	27.0	29.5	0.0	277.0
1 5 Non-Rotary Power Saw	0.7	0.0	0.0	9.9	0.0	0.0	0.0	0.0	1.4	0.0	12.0
Total Hours	124.8	1.7	145.2	538.6	15.8	353.4	447.1	311.8	46.8	817.5	2 803.8
2 0 Drill Work Drill Press	111.7	0.1	151.0	407.7	8.2	253.4	303.6	176.7	119.9	460.1	1 992.4
2 1 Boring Mill	0.0	0.0	10.8	139.2	0.3	207.7	326.2	345.9	100.3	473.1	1 603.5
2 4 Boring Mill	0.0	0.0	0.0	1.1	0.0	0.0	0.0	0.0	0.0	0.0	1.1
Total Hours	111.7	0.1	161.8	548.0	8.5	461.3	629.8	522.8	220.2	933.2	3 597.4
3 0 Manual Work Bench Work	247.0	0.4	192.3	564.2	0.0	378.8	488.7	416.5	141.8	2 399.9	4 856.6
3 1 Manual Work Assembly Work	11.1	0.0	22.1	113.2	1.3	151.2	226.8	177.8	170.3	6 746.8	7 620.6
3 2 Manual Work Marking up Work	27.0	0.0	30.8	123.0	0.2	122.1	163.3	226.4	57.0	651.6	1 401.4
3 4 Manual Work Other Manual Work	16.2	0.0	5.7	41.5	0.0	68.1	42.7	0.3	0.0	0.0	174.5
Total Hours	301.3	0.4	290.9	841.9	1.5	720.2	921.5	821.0	369.1	9 798.3	14 066.1
4 0 Surface Work Chamfering	4.1	0.0	6.5	40.8	0.0	25.3	27.0	49.1	11.9	93.3	258.0
4 1 Surface Work Cleaning	80.9	0.0	19.8	7.7	0.0	16.2	15.6	9.0	0.0	131.6	199.9
4 2 Surface Work Metal Finishes	12.6	0.0	5.9	22.3	0.0	24.8	87.2	0.1	0.0	72.0	224.9
4 3 Surface Work Non-Metal Finishes	2.5	0.0	11.7	26.7	0.0	34.6	32.2	16.4	12.4	196.2	332.7
Total Hours	100.1	0.0	43.9	97.5	0.0	100.9	162.0	74.6	14.3	493.1	1 096.4
Sum of Total Hours	1 198.0	575.4	1 137.0	2 921.8	952.3	1 664.5	2 324.7	2 033.0	805.1	13 104.8	26 716.6

Operational Code 1st and 2nd Digit

Machining Type — Machining Hrs. in Respect of Dimensional Class and Machining Type

Fig. 4.12. Attachment of Parts of a Diesel Engine to the Individual Machining Types

2. Digit Machining Type

3. Digit Machining Process

4. Digit Dimensional Classes L

5. Digit Dimensional Classes D

0	Automatic Lathe Work without Threading

	3. Digit Machining Process
0	Automatic Lathe Bar Work
1	Chucking Automatic Work
2	Turret Lathes for Bar Work
3	Turret Lathes for Chucking Work
4	Drum–Type Turret Lathe for Bar Work
5	Drum–Type Turret Lathe Chucking Work
6	Multi–Spindle Automatic Work
8	Other Automatic Lathe Work
	Vacant

1	Lathe Work without Threading

0	Horizontal, Centre Lathe Work
1	Vertical Turning and Turning Mills, Facing
2	Eccentric, Oval, Relief, Profile Lathe Work
3	Horizontal Copying Lathe Work
4	Vertical Copying Lathe Work
5	Form Turning
6	Welding Joint Turning Work
8	Single–Purpose Lathe Work
9	Other Lathe Work

2	Grinding

0	External Grinding
1	Internal Grinding
2	Centreless Grinding
3	Copying Grinding
6	Honing, Lapping (External)
7	Honing, Lapping (Internal)
8	Single Purpose Grinding Work
9	Other Grinding Work

4. Digit Dimensional Classes L

L	
16	A
20	B
25	C
32	D
40	E
50	F
63	G
80	H
100	I
120	J
160	K
200	L
250	M
320	N
400	Ø
500	P
630	Q
800	R
1000	S
1200	T
1600	U
2000	V
2500	W
3200	X
4000	Y
< 4000	Z

5. Digit Dimensional Classes D

D	
16	A
20	B
25	C
32	D
40	E
50	F
63	G
80	H
100	I
120	J
160	K
200	L
250	M
320	N
400	Ø
500	P
630	Q
800	R
1000	S
1200	T
1600	U
2000	V
2500	W
3200	X
4000	Y
< 4000	Z

Fig. 4.13. Operational Code 3rd Digit Machining Process (Rotary Machining Area)

Fig. 4.14. Formation of the Workpiece Supplementary Code

1. Digit — Coarse Dimension (mm)

Code	L (A)	D (B)
0	100	≤ 100
1	100	≤ 400
2	400	≤ 100
3	400	≤ 400
4	1000	≤ 100
5	1000	≤ 400
6	1000	≤ 1000
7	1600	≤ 1000
8	1600	≤ 1600
9	> 1600	> 1600

2. Digit — Weight (1 to = 1000 kg)

Code	Weight
0	≤ 2 kg
1	> 2 kg < 20 kg
2	> 20 kg < 200 kg
3	> 200 kg < 750 kg
4	> 750 kg < 2,5 to
5	> 2,5 to < 5 to
6	> 5 to < 10 to
7	> 10 to < 25 to
8	> 25 to < 40 to
9	> 40 to

3. Digit — Machining Characteristics

Code	Machining Characteristics
0	Chip Cutting
1	Cutting Several Components
2	Cutting and Non-Cutting Op.
3	Cutting and Joining Operations
4	Cutting Non-Cutting and Joining Operations
5	Non-Cutting Operations
6	Forming and Joining Operations
7	Joining Operations
8	Assembly
9	Others

4. Digit — Material

Code	Material
0	Steel — Non Alloy and < 5 % Alloying Elements — Without Heat-treatment
1	Steel — Non Alloy and < 5 % Alloying Elements — With Heat-treatment
2	Steel — Alloy Steel > 5 % — Without Heat-treatment
3	Steel — Alloy Steel > 5 % — With Heat-treatment
4	Cast Iron No Alloying Elements
5	Alloy Cast Iron Malleable Iron
6	Cu and Cu – Alloys
7	Al and Al – Alloys
8	Other Materials
9	Several Materials

5. Digit — Initial Form

Code	Initial Form
0	Casting
1	Forging
2	Bar
3	Pipe
4	Ingot
5	Sheet Metal up to 12 mm
6	Sheet Metal over 12 mm
7	Stock Parts
8	Other Forms
9	Several Forms

A total of 403 machining processes and activities are covered by the present system. The operational code has been tested on over 80000 operations. This classification stage constitutes the link between the general and the specific planning tasks. It enables the use of an independent cost-centre identification which is of decisive importance for the practical introduction. If a cost-centre includes various machining processes, it can be sub-divided into planning groups corresponding to the process level. The same procedure may also be adopted for the dimensional fields which corresponds to the working areas of the respective equipment in this planning group. These measures enable the production control to tie together the operational task with the equipment and plant available. It should be noted that the so-called reserve places, previously provided for in the operational chart, are no longer required. On the contrary, they are now covered by the production control which shows the work distributor the alternative possibilities. They have only to be checked to ascertain whether the specific preliminary conditions exist.

In testing the application possibilities of the two partial systems for diverse planning tasks, it was found that additional planning characteristics were required for investigations at workpiece level. They have to be covered by a special workpiece supplementary code. This five-figured code supplements the process-independent workpiece code by classifying the coarse dimension, weight, machining characteristics, material and initial form into ten classes each (Fig. 4.14). Special consideration was given to the problems of technical investment and layout planning. The digit of the coarse dimension class constitutes a concentration of the two main dimensions. The limits of the weight classes are atuned to the crane load. The code is a constituent part of the operational code. The inter-play between the two code components is shown in Fig. 4.15. It comprises the operational code used to describe the individual operations and work-

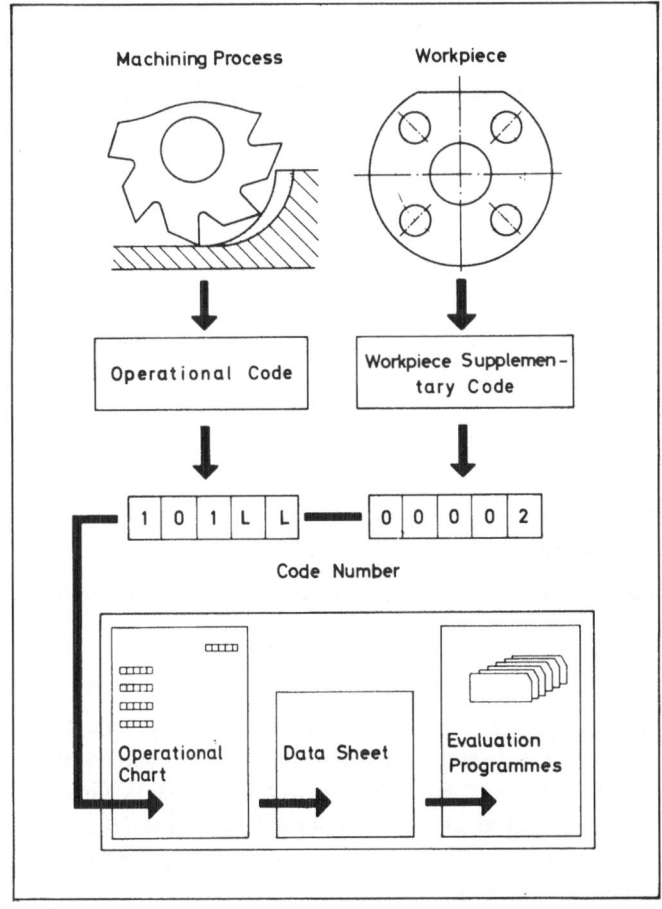

Fig. 4.15.
Sub-System 'Operation'

piece supplementary code with additional machining-orientated characteristics. The operational chart is the data carrier because the coding is carried out in connection with the process planning. The data sheet contains all the coding data and serves for the computer input. The evaluation programmes are used for the various investigation objectives. This sub-system constitutes the transition from the design stage to machining considerations. It principally serves the technical investment and layout planning and also as a basis for the build-up of machining families in the process planning.

4.2.3 Sub-System 'Equipment'

The overall term 'Equipment' has been chosen in a manner to take the wide spectrum of different plant equipment into consideration, and to provide for subsequent extension.

Fig. 4.16 shows the equipment structure. The area 'Manufacturing Equipment' represents a first extension stage within the framework of the technological structure planning with the crucial points of the machine tools and plant as well as the manual work and assembly places.

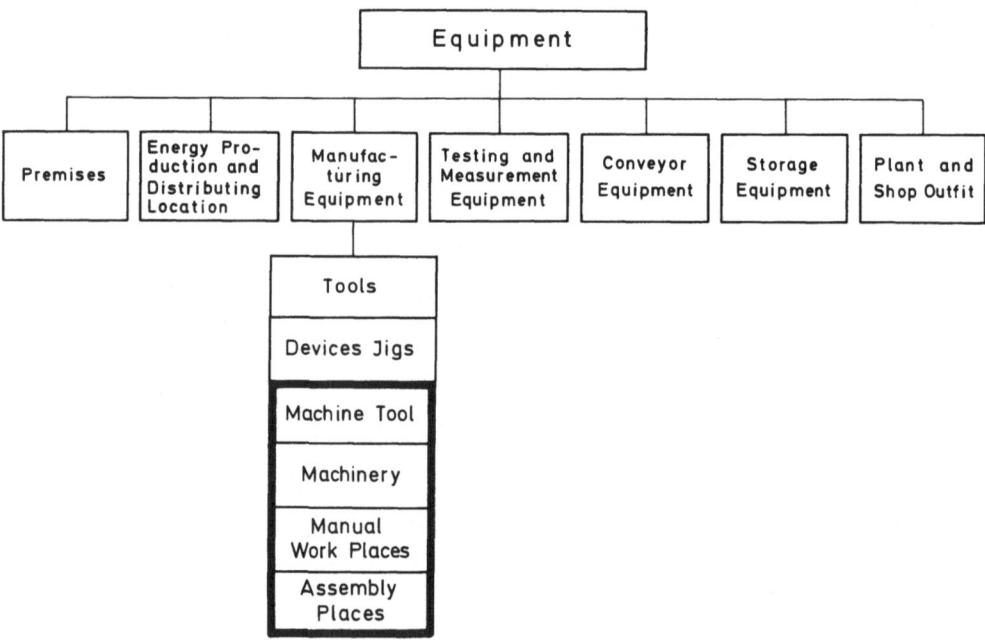

Fig. 4.16. Manufacturing Equipment – First Extension – up Stage within the Framework of the Technological Structure Planning

The 9-digit equipment code is used as a working aid to determine the equipment structure. It consists of a 4-digit equipment code and a 5-digit equipment supplementary code. Fig. 4.17 provides an example of the 4-figure equipment code based on Kindhauser principles [28]. The 5-figured equipment supplementary code (Fig. 4.18 and 4.19) describes supplementary data for investment and layout planning, and represents the analogue part of the workpiece supplementary code in the operational systematic.

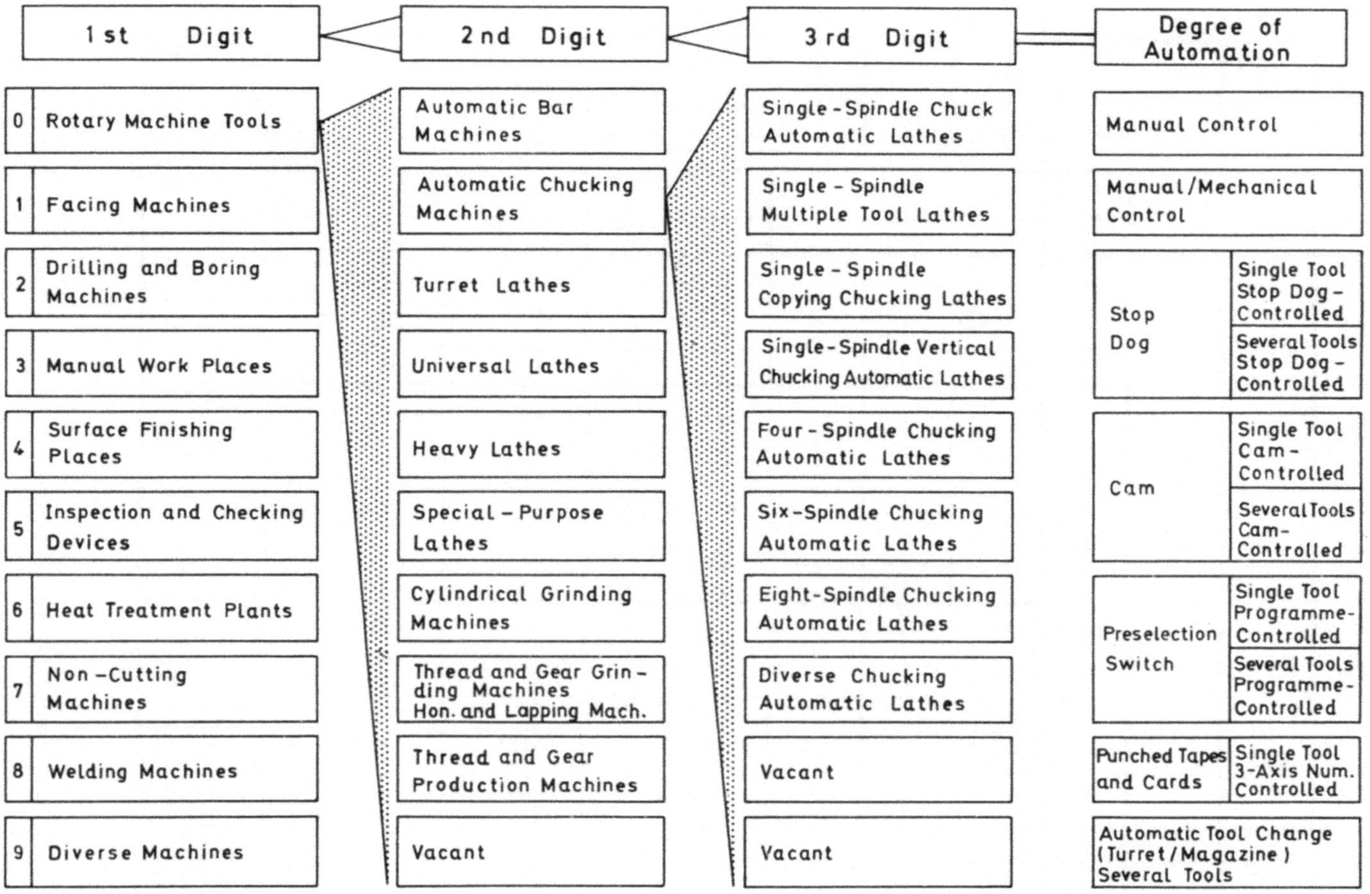

	1st Digit	2nd Digit	3rd Digit	Degree of Automation
0	Rotary Machine Tools	Automatic Bar Machines	Single-Spindle Chuck Automatic Lathes	Manual Control
1	Facing Machines	Automatic Chucking Machines	Single-Spindle Multiple Tool Lathes	Manual/Mechanical Control
2	Drilling and Boring Machines	Turret Lathes	Single-Spindle Copying Chucking Lathes	Stop Dog — Single Tool Stop Dog-Controlled
3	Manual Work Places	Universal Lathes	Single-Spindle Vertical Chucking Automatic Lathes	Stop Dog — Several Tools Stop Dog-Controlled
4	Surface Finishing Places	Heavy Lathes	Four-Spindle Chucking Automatic Lathes	Cam — Single Tool Cam-Controlled
5	Inspection and Checking Devices	Special-Purpose Lathes	Six-Spindle Chucking Automatic Lathes	Cam — Several Tools Cam-Controlled
6	Heat Treatment Plants	Cylindrical Grinding Machines	Eight-Spindle Chucking Automatic Lathes	Preselection Switch — Single Tool Programme Controlled
7	Non-Cutting Machines	Thread and Gear Grinding Machines Hon. and Lapping Mach.	Diverse Chucking Automatic Lathes	Preselection Switch — Several Tools Programme-Controlled
8	Welding Machines	Thread and Gear Production Machines	Vacant	Punched Tapes and Cards — Single Tool 3-Axis Num. Controlled
9	Diverse Machines	Vacant	Vacant	Automatic Tool Change (Turret/Magazine) Several Tools

Fig. 4.17. Formation of the Equipment Code for Machine Tools and Plants

41

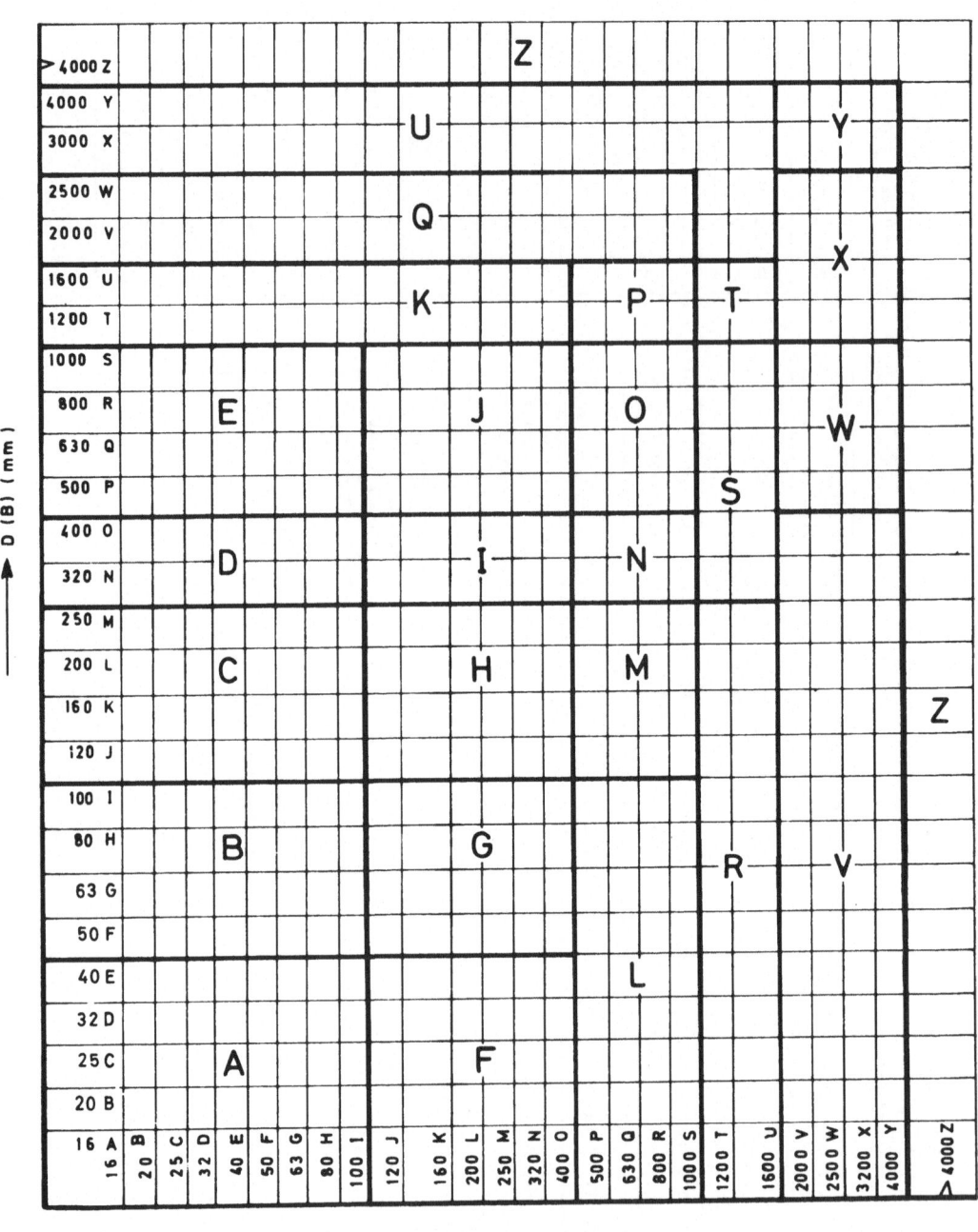

Fig. 4.18. Equipment Supplementary Code lst Digit

The equipment supplementary code provides information concerning a concentrated dimension class for preliminary studies as well as machine power, the year of delivery and the prices. The equipment catalogues are used as a basis. For the input to the computer a special equipment data collection sheet is provided.

	2.Digit	3.Digit	4.and 5.Digit
	Machine Power	Price	Year of Delivery
	kW	Sfr.*	
A	1.6	$10^3 x$ 1	
B	2	1.6	
C	2.5	2.5	
D	3.2	4	
E	4	6.3	
F	5	10	
G	6,3	16	
H	8	25	
I	10	40	
J	12	63	
K	16	100	2 Digits
L	20	160	
M	25	250	
N	32	400	
O	40	630	
P	50	$10^6 x$ 1	
Q	63	1.6	
R	80	2.5	
S	100	4	
T	120	6.3	
U	160	10	
V	200	16	
W	250	25	
X	320	40	
Y	400	63	
Z	> 400	> 63	

* Sfr. = Swiss Frank

Fig. 4.19. Equipment Supplementary Code 2nd - 5th Digit

Fig. 4.20 shows the equipment data sheet. This sheet contains the data needed for the general planning area. The equipment data collection sheet was tested in a workshop with a number of 2000 equipments. The most difficult task thereby was to convince the various planning sections of using uniformly coded data. For the collection of further data, specific follow-up cards are provided, which are necessary for the assignment of the operational tasks to the equipment within the framework of the process planning. In an analogous manner, further follow-up cards can be developed for the layout planning, maintenance of NC-machine tools and other special problems according to the various crucial points of rationalization.

The 5-figure operational code constitutes the connection between the operational tasks and the equipment, and describes the individual processes of the machine tools and plants. The combination of the two sub-systems allows a comparison of the equipment at disposal and the machining structure of a certain production programme (Fig. 4.21).

The machining centre where various individual processes are grouped together constitutes a special case. At coding, the equipment has to be individually determined for each process. This procedure enables us to code complete manufacturing systems and reduce them to an overall planning level.

43

Equipment Data Collection												
	Object Level										Process Level	
Sheet	Producer	Designation of Equipment	Identification Number	Manufacturing System	Cost Centre Planning Group Work Place Number	Equipment Code Number	Equipment Supplementary Code Number			Number of Processes	Operational Code Number	

Follow-up Data for:

– Layout Planning
– Process Planning
– NC-Programming

Fig. 4.20. Equipment Data Collection Sheet

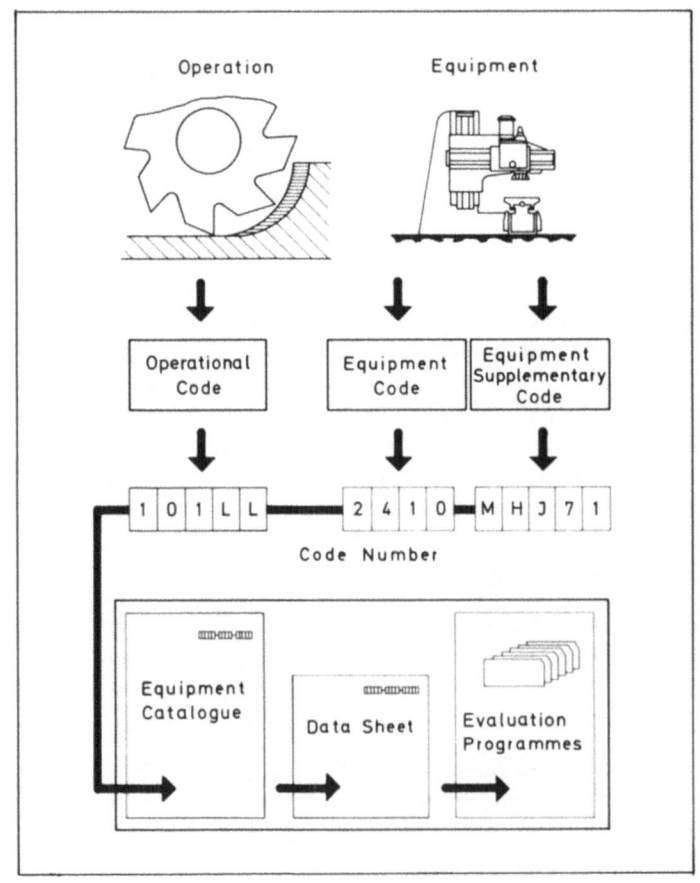

Fig. 4.21. Sub-System 'Equipment'

4.3 Application of the Structure Planning

Dependent on the objective, the sub-system may be used individually or collectively to solve the different planning tasks. The area of statement of the system may also be adapted to the respective requirements with the aid of additional data. In most cases, however, the general planning data is sufficient. Even if there is a possibility of using computer data directly for the individual planning tasks, experience shows that in many cases a conclusive and economic basis can be created for preliminary studies with the aid of reference number sheets (Fig. 4.9, 4.12, and 4.13). These sheets can be brought up to date at certain intervals and adapted to met the increasing requirements.

4.4 Technological Data Bank

If the tasks are more complex, e.g. within the framework of investment and layout planning, as well as in the case of standardization efforts in the design and process planning, it is of great advantage for the practical application if various basic variants can be exercised in order to obtain a conclusive planning basis. This is made particularly clear with an example drawn from investment planning (see Section 5.2), where from the market trend viewpoint a probable or a max/min. programme variant is adopted with a view to determining the effect on the production hours.

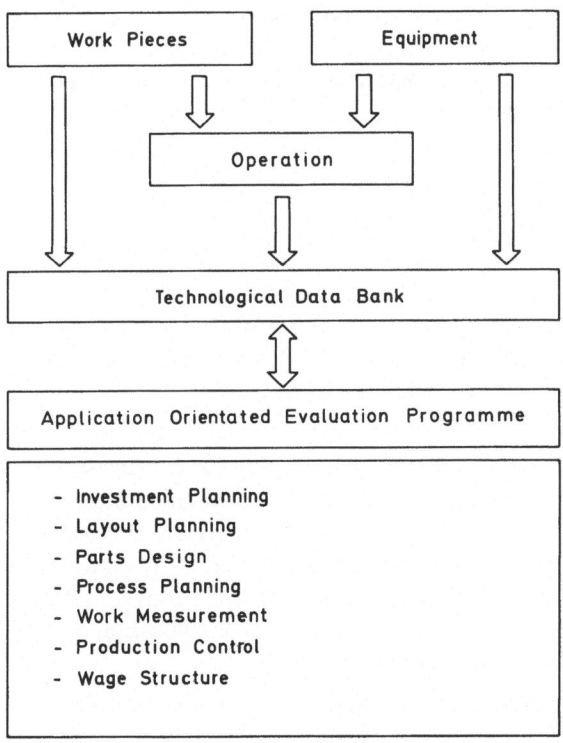

Fig. 4.22. Technological Data Bank and Evaluation Programme

In such cases, representative product types stored, in a data bank in a coded form in accordance with the fundamental principles of the classification system, serve as the basis for investigation. With the aid of application-orientated evaluation programmes, variants can then be calculated according to various aspects. In doing so, the basic principle to be observed is that the evaluation programme should not offer a tailormade starting position, but should be preferably laid out in a universal manner, and any adaptions, e.g. the hourly requirement for additional work and spares, are to be estimated by an experienced specialist.

The investigations show quite clearly that with large quantities of data the elaboration of variants would be very time-consuming without the use of a computer. The elaboration of a planning basis for the laying out of diesel engine manufacture, for which a number of variants have to be calculated, the preparatory work took about one week, whereas six man/months were required for the manual evaluation.

Fig. 4.22 illustrates the technological data bank scheme in which the workpiece, operations and equipment data are shown in a coded form. The data can be called up, sorted and calculated with an application-orientated evaluation programme.

4.5 Evaluation Programme for GT

The structure of an evaluation programme is described with an example of similarity analysis for the creation of parts groups. The entire procedure may be compared to a sieving operation. In the first stage, the mesh is very coarse and then it becomes gradually finer through the use of other similarity characteristics. The rough sorting is effected with the aid of standardzed similarity fields via

— Shape description,
— Type of material,
— Machining process,
— Dimensional areas.

The individual similarity fields represents so-called 'macros', in which the various classification characteristics are grouped together. Depending on the kind of problems, the similarity fields can then be grouped together to form an application-orientated investigation order. The great advantage of this procedure is that the problem can be limited with a very small amount of input data without losing sight of the overall problem.

In a second stage, the fields can be divided into their characteristics and a fine sorting operation carried out. This procedure facilitates a quick and objective-orientated analysis of the crucial points of the investigation, and reduces considerably the number of computer throughputs. With the aid of similarity fields, a comprehensive planning basis can be created in which a limitation of the overall machining, e.g. turning, milling and drilling, can be effected during the same sorting operation for the selection of the parts groups. This is very advantageous for layout planning.

The coded representative product types in the technological data bank constitute the starting point. The overall parts spectrum is not considered completely because the investigations showed that a relatively small number of deviation errors occur with statistically selected product types. For that reason the coding effort can be reduced considerably. This partial problem is described with the aid of the example of an investment planning for a pump factory in chapter 5.

The following briefly describes the individual structure elements of the evaluation programme. Fig. 4.23 shows the programme build-up. The technological data bank with the coded classification characteristics of the representative product types is the starting-point.

In a first step the individual product types with technological similarity are collected within the criteria of, e.g.

— Machining characteristics,
— Batch characteristics,
— Accuracy requirements,
— Inspection requirements.

In this stage, the overall aspects of the corporate organization stand in the forefront. Measures are taken to ensure at first that apparently similar part groups, which are essentially different from the manufacturing technique viewpoint, are not conglomerating, e.g. the gears of locomotives and weaving machines.

In the evaluation programme the similar product types are collected together in product groups. By that, they form a first level in the similarity structure. In the evaluation

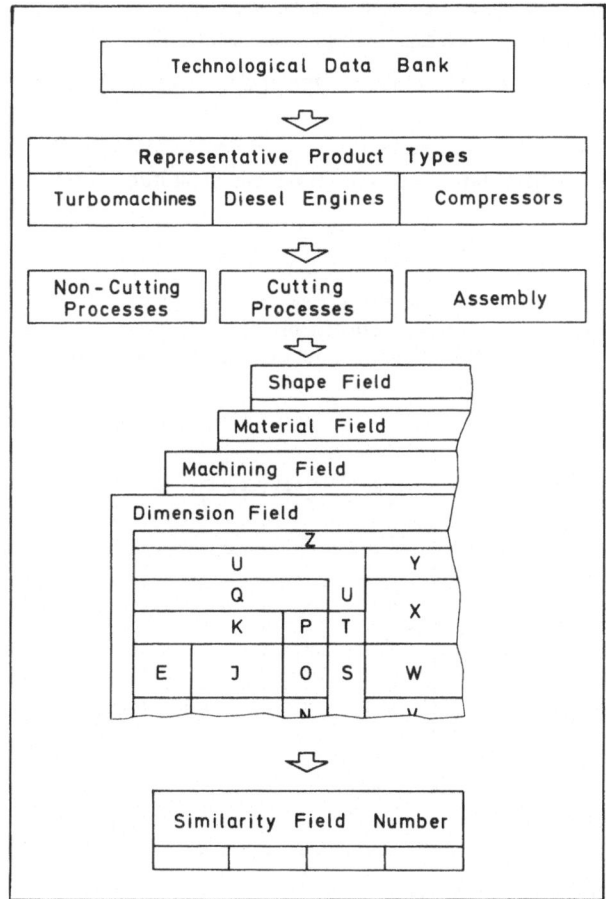

Fig. 4.23. Evaluation Programme
– Similarity Analysis

programme, the individual product types are provided with a product code. For the formation of these product groups, apart from the similarity criteria accounted for, further corporate influencing factors have to be taken into consideration, e.g. situation of labor market, existing plant equipment and market political considerations.

In the second step, the structuring within the manufacturing technical criteria is effected with the aid of the workpiece and operational code, e.g.

— Cutting processing,
— Non-cutting processing,
— Assembly.

In the workpiece and operational code as a result of these considerations, the non-cutting processing is also included with a somewhat reduced conclusiveness. When a differentiated planning base is required within the non-cutting processing, it is necessary to insert supplementary code for workpiece such as those used for sheet metal parts. Attention is drawn to the investigation carried out at the University of Aachen [29]. In this investigation the crucial point lies in the cutting process.

The third step provides for a grouping of parts machining according to defined similarity fields. These represent a first standardization in respect of GT and it serves as a basis for layout planning and the build-up design and machining families. For the determination of the similarity structure in the parts machining of the individual product groups, the following main criteria have been taken as a basis, whereby, according to detailed investigations, the following sequence has been proved as suitable:

— Collection of similar main criteria of the workpiece shape and machining characteristic within the shape fields.
— Collection of material classes with the same machining requirements in material fields. This criteria is decisive for the characteristic of the parts machining and the operation sequence.

Example of the Shape Field 00 :				Cutting Processes – Rotary Parts without Deviation Disc–Shaped and Flanges without Thread				
Digits of Workpiece Code	Geometric Shape			Branch	Machining Characteristics			
Position	Main Shape 1	Secondary Shape 2	Thread 3	Part Class 4	External 5	Internal 6	Facing Gear Cutting 7	Secondary Drilling 8
0		x	x	x	x	x	x	x
1		x			x	x	x	x
2		x			x	x	x	x
3	x	x			x	x		x
4		x			x	x	x	
5		x			x	x	x	x
6		x						x
7	x	x					x	x
8		x					x	
9								x

Fig. 4.24. Similarity Field, 1st and 2nd Digit, Shape Field

48

– Formation of machining fields for the parts machining in relation to the preceding shape and material fields.
– Formation of dimensional fields based on the frequency analysis of the parts spectrum.

The limits of shape, material, machining and dimensional fields should be fixed as wide as possible. The aim of this requirement is to create a general starting basis. This should serve the process planning as well as the general tasks, e.g. as a basis for the technical investment planning. With the help of an evaluation programme, the workpieces with the same shape, material, machining and dimensional fields should be collected together in so-called 'similarity fields'. The resulting code number combination represents a coarse description of the parts machining. It can be called a planning factor, with whose help it is possible to describe a similarity structure of a determined parts spectrum in a comprehensive way.

The following example of rotary machining illustrates the structure of the similarity fields.

Fig. 4.24 shows an example of the workpiece similarity in respect of the geometric shape and the machining characteristic as a so-called 'shape field' with the aid of an assignment table.

The first digit of the shape field is made from 10 part classes of the workpiece code (Fig. 4.25). The second digit includes the characteristic combination according to the corresponding assignment table. Theoretically the system may have up to 100 shape

Part Classes				
0		Rotary Parts without Deviation		
1		Rotary Parts		General
2	Forming by Cutting Processes	with Deviation		Casing Housing
3		Non - Rotary		General
4		Parts		Casing Housing
5		Forging Pressing Stamping etc.	= 3	General
6			= 4	Casing Housing
7	Forming by Non-Cutting Processes	Casting Sintering etc.	= 3	General
8			= 4	Casing Housing
9	Assembly Groups			

Fig. 4.25. Similarity Field – 1st Digit – Corresponding with the Part Classes of the Workpiece Code

fields. However, for the purpose of this investigations 21 fields have been defined for the cutting processes.

10 material classes of the workpiece supplementary code are forming the basis for the material fields (Fig. 4.26). In doing so, a condensation to five material fields was established. In respect of the machining process the heat treatment provides an important criteria.

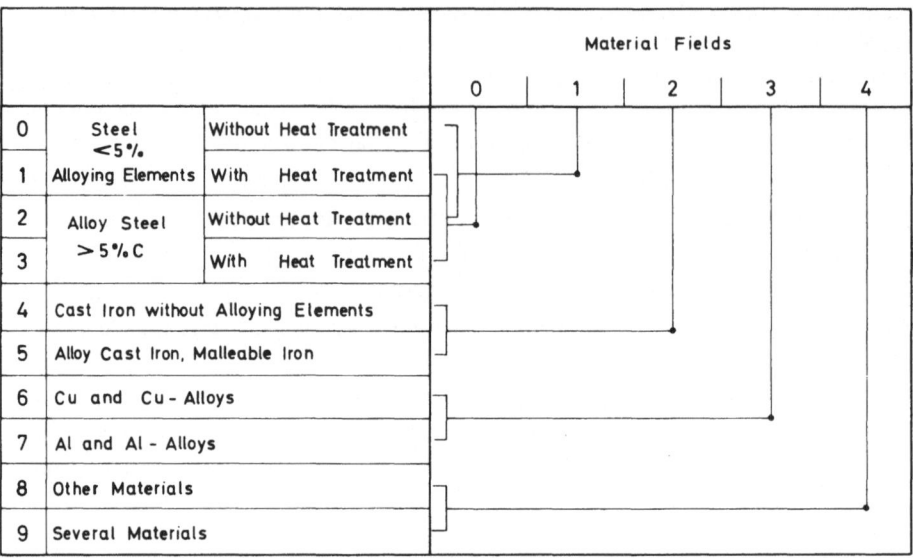

Fig. 4.26. Similarity Field – 3rd Digit – Material Fields Represent a Combination of the Material Fields Represent a Combination of the Material Classes Based on the Workpiece Supplementary Code

Operational Code Machining Types	Shape Field						No.		00		Material Field			No.		1 – 4		
	Machining Fields																	
	01	02	03	04	05	06	07	08	09	10	11	12	13	14	15	16	17	18
Turning	X	X	X	X	X	X	X	X	X	X	X	X	X	X	X			
Grinding		X	X	X	X	X				X	X			X				
Milling			X	X	X	X						X	X	X	X			
Shaping / Slotting				X	X	X	X							X				
Broaching, Keyway Cutt.					X	X		X		X	X							
Drilling						X			X		X	X		X				
Benchwork, Assembly			X	X	X	X	X	X	X	X	X	X	X	X	X			

Fig. 4.27. Similarity Field – 4th and 5th Digit – Machining Fields Build-up on the Operational Code

On the basis of the shape and material fields, the machining possibilities in the sense of a first standardization, are established. The machining combinations are shown as so-called 'machining fields'. Using Fig. 4.24 as an example Fig. 4.27 provides an assignment table for machining fields.

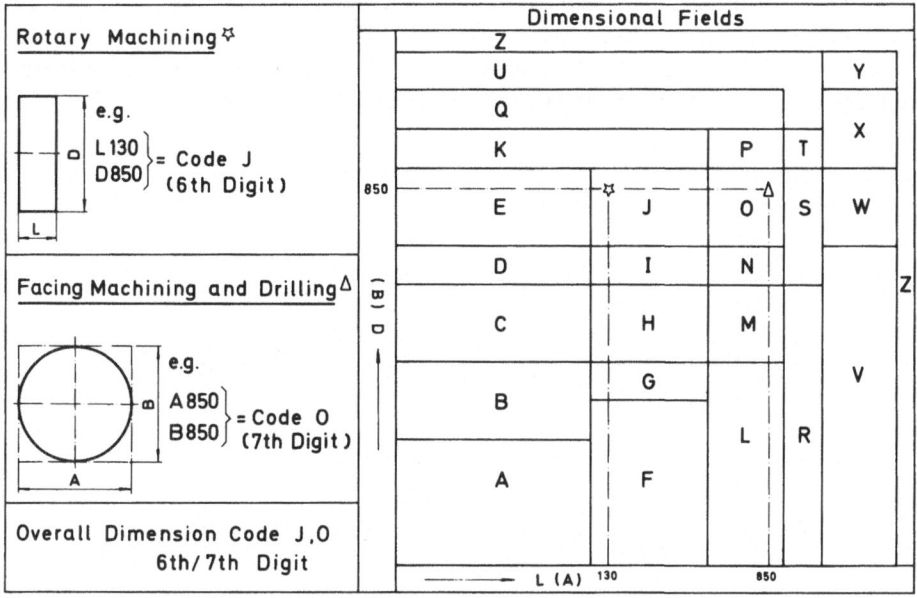

Fig. 4.28. Similarity Fields – 6th and 7th Digit – Dimensional Fields

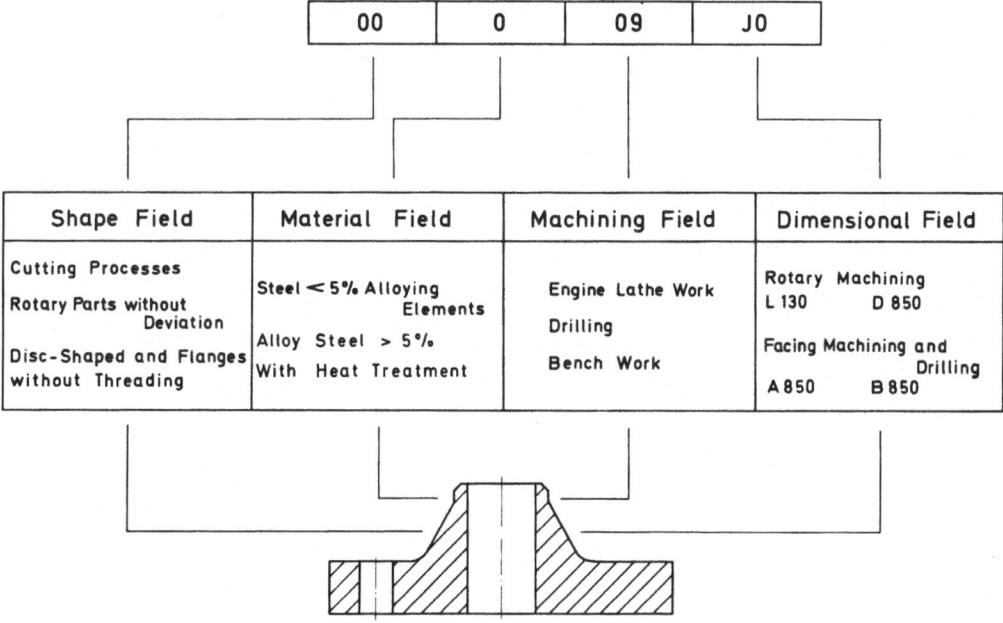

Fig. 4.29. Similarity Fields, Coding Example

The dimensional fields Fig. 4.28 are based on the fundamental grid (26 classes) of the operational code, whereby the workpiece is coded in accordance with rotary, facing or other machining possibilities.

Fig. 4.29, a rotary part, is used as an example for the combination of the individually described fields within a similarity field.

The structure for the establishment of the similarity fields which can be made with the help of an evaluation programme is summarized in Fig. 4.30. Depending on the production structure of the corporation, the similarity fields can be adapted to the respective characteristic conditions by changing the combinations of characteristics.

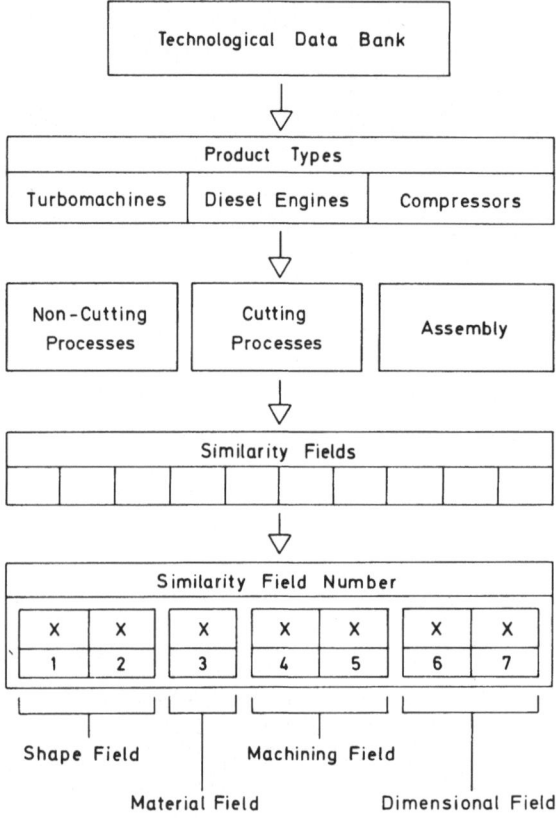

Fig. 4.30. Similarity Fields of the Evaluation Programme

5 Investment and Layout Planning

Within the framework of GT, investment and layout planning is used to determine the optimum rationalization stage and to lay out the GT-manufacturing systems for a selected parts spectrum. Using the general principles of investment and layout planning as a basis, a procedure is recommended for this problem statement which is marked by its methodical and flexible application.

In a simplified manner, the investment planning can be described as a process which allocates a suitable equipment spectrum to a parts spectrum to be manufactured. The investment planning within the framework of corporate planning is based on the provision of the equipment demand. A preliminary condition thereby is that the product programme is firstly established by the long-term objectives of the corporation. Using the market planning as a basis, the production programme will then be derived, corresponding to the market demand. With the aid of this basic information and knowledge of the product requirements and technology, a systematic technical planning process is started which will finally provide information concerning the necessary equipment.

As a result of a subsequent comparison with the existing capacities and an economic investigation, which duly considers the corporate objectives, it is possible to determine the necessary measures to be taken in respect of investment, rationalization and displacement.

Layout planning deals with the solution to the factory-related group of problems, which include the actual layout and creation of GT-manufacturing systems.

The relationship between the investment and layout planning is shown more clearly in Fig. 5.1. Although within the layout planning, we differentiate between new and rearrangement planning, there is very little difference in the actual procedure. Rearrangement planning has to consider the existing equipment and machines, whereas new planning is, in addition, charged with the procurement of new machines.

The product and production programme provides the most important initial information for both types of planning. The term 'product programme' infers all the products and parts that a corporation has to offer. This also includes all the products and parts that are in the stage of development, but will be incorporated in the product programme at an early date, and therefore have an influence during the planning period.

The production programme represents the quantity aspects of the product programme — in relation to a planning period — and thus provides us with information as to the overall extent of self-manufacture. The stepwise procedure employed to determine the optimum layout forms represents the crucial point of the working method used in layout planning for GT-manufacturing systems (Fig. 5.2).

First, groups of parts with similar machining characteristics, which also ensure an economic loading of the plant and equipment, are formed from the parts spectrum to be investigated. The first step involves analysing each of these machining groups for the possibility of applying a GT-line. If the analysis does not lead to a sufficient loading,

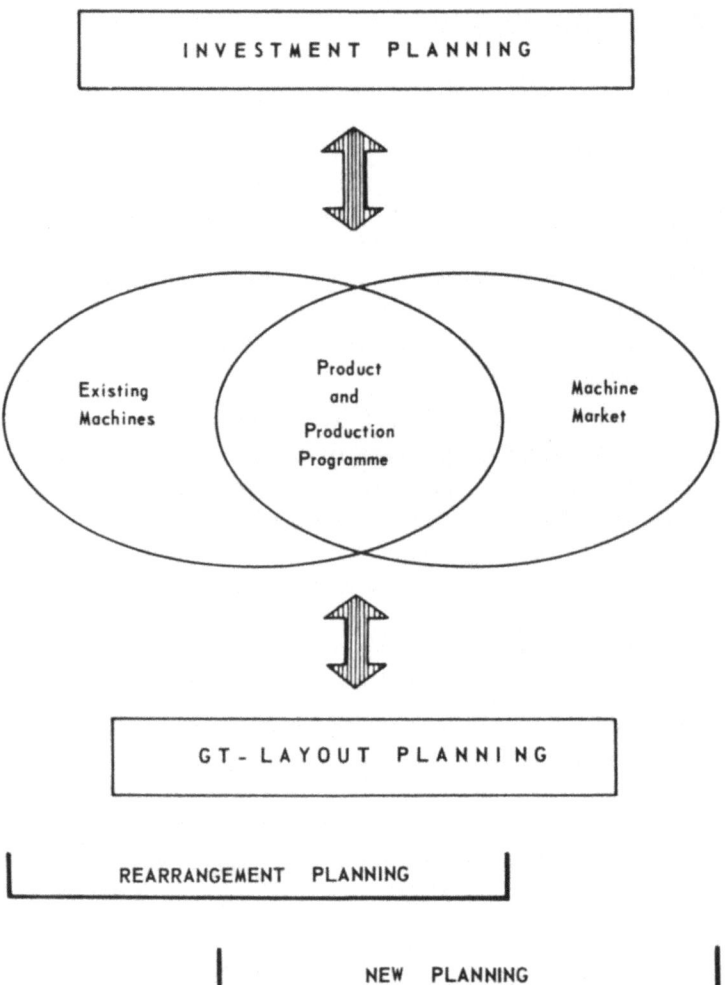

Fig. 5.1. Relationship between the Investment and Layout Planning

the same machining groups are examined again in respect of the second rationalization stage, namely a GT-cell. Should the result again prove to be negative, the machining groups are finally examined for the application of GT-centres. If the first phase should result in an insufficient loading, then it is possible to create a better-loaded starting position by extending the machining group, whereby a lower degree of similarity will have to be accepted. All remaining parts that cannot be allocated to a form of GT-layout constitute the starting position for the planning of a functional layout.

Whereas a study of the literature shows that a great number of individual problems have been dealt with in the field of general technical investment planning, only little work has been devoted to the comprehensive interrelationships [2, 30, 31, 32].

A multitude of investigations have been carried out in connection with the economic considerations of investment planning and provide mainly information about individual investments and the most economic application of the capital for specified variants.

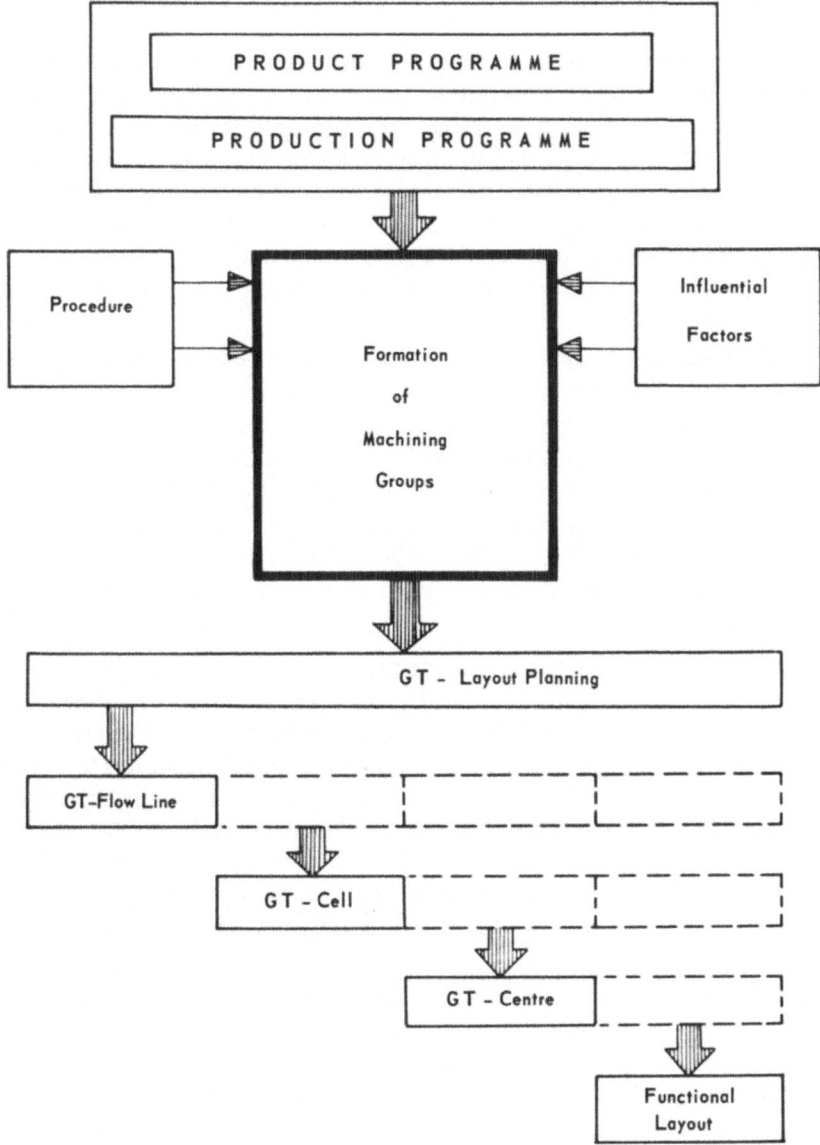

Fig. 5.2. Layout Planning Stages

With GT-manufacturing systems, the economic consideration is mainly concentrated on the reduction of the total throughput time and the storage as well as simplification of the work organization. In the main, these factors are difficult to determine with the normal cost accounting methods.

To reach comprehensive investment decisions, corresponding decision aids have still to be developed on the accounting management side. The following is devoted to the individual steps of the concrete procedure and duly considers the development work already mentioned in Chapter 1.

5.1 Investment Planning within the Framework of GT

As an integral part of corporate planning, investment planning constitutes a basis which enables us to react to external market-conditions. Apart from the resultant future parts demands and the parts characteristic, knowledge of the probable deviations caused by external uncertainty factors of the market also plays a significant role in the layout of manufacturing systems. On the one hand, these are determined by the future market fluctuations, which effect the total number of hours, and on the other hand conditioned within the planned market quantitiy by the deviations from the assumed market mixture. The representation and the determination of these deviation factors will be investigated with the aid of a model.

The reliability of this planning basis is directly dependent on the selected dispositions and input data, which are of decided importance and difficult to obtain in practice. Using this situation as a starting point, a special procedure has been developed, which is also to be employed for the subsequent layout planning, for the execution of the investment planning and in particular for a rational preparation of the decision data. The input data and planning procedures are shown in Fig. 5.3. Depending on the required conclusiveness and the degree of detailing, three stages are provided — rough I, mean II and fine III. These stages correspond to the classification system structure for the coding of the working operations (Chapter 4, Fig. 4.8).

The procedure is characterized by the columns 'demand' and 'offer'. The product programme, — as a starting basis — is shown on the demand side and has the aim of determining the trend pattern of future manufacturing capacities for the individual processes. A relatively difficult problem thereby is to convert the trend, according to design parameters, into specified manufacturing forecasts. A closer examination reveals, however, that there is no direct relationship between the design and manufacturing trend in a large number of cases. For example, the increasing diameter of a certain product type does not lead necessarily to a change in the hourly demand or machine sizes.

Fig. 5.4 shows clearly again that there is no necessity for a relationship between the parts and machining structures in a large number of cases. In case I, two pump types are shown which are very similar in parts structures, but have different machining structures. In case II, however, we find the reverse position for three other pump types. They have a relatively very different parts structure and very similar machining structures. But now back to Fig. 5.3.

The assumed maximum and minimum market trend for the next planning period is determined for every product type included in the product programme. Within this scatter area and with the aid of the production programme, the planning factor is laid down from which the manufacturing structure can be derived.

As a result of a thorough analysis of the planning tasks, it is found that 2/3 of the total effort is needed for the collection and evaluation of the basic data and only 1/3 for the real planning activity. On the other hand, and particularly with complex problems, a number of variants should be worked out if we are to offer reliable decision fundamentals.

With a view to reducing the planning effort to an absolute minimum, the investigations are generally made with the aid of a representative types spectrum of products. This procedure, which represents an essential rationalization, is therefore of particular importance.

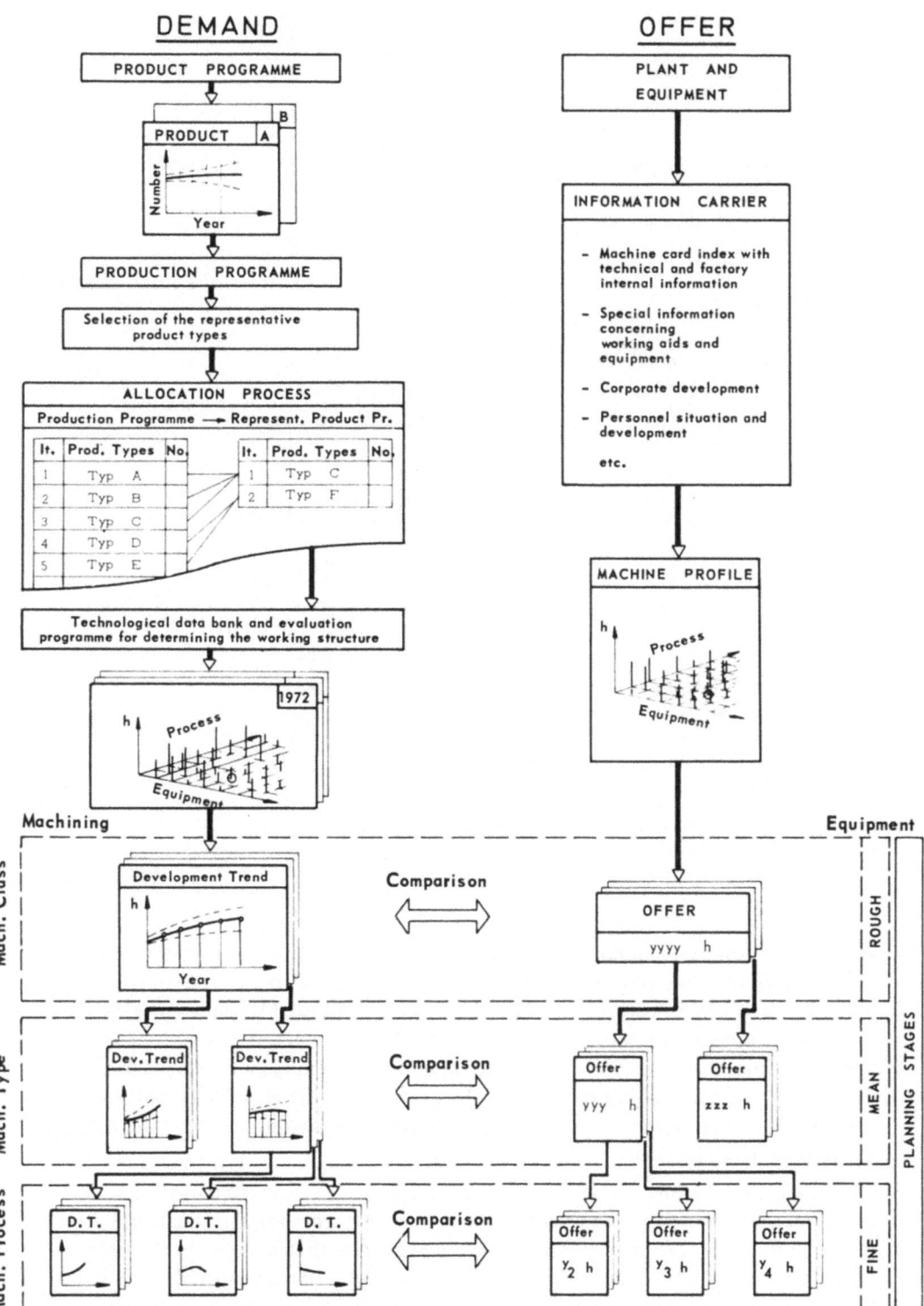

Fig. 5.3. Procedure Model for Investment Planning

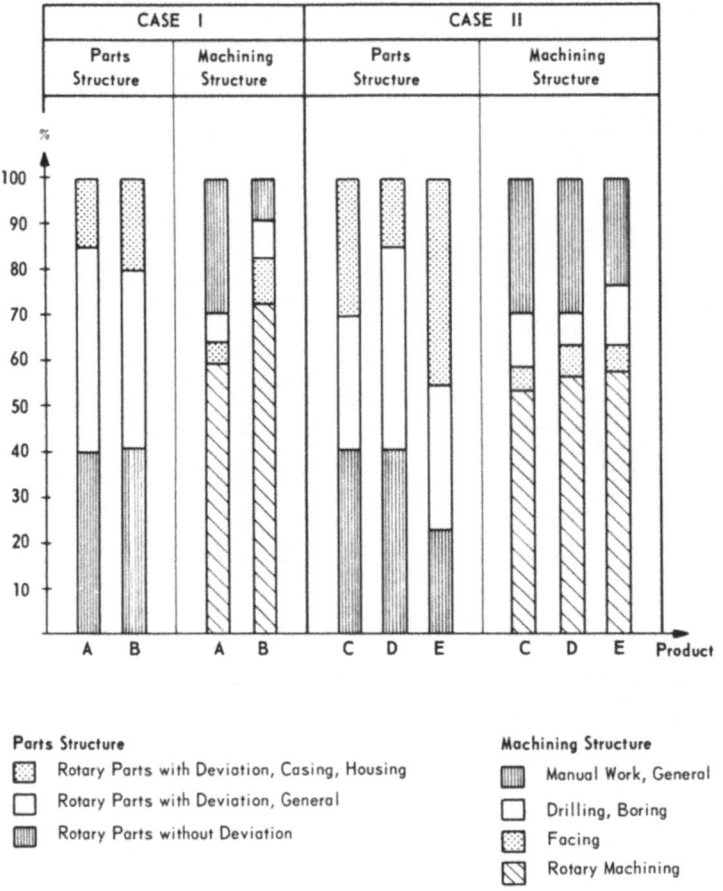

Fig. 5.4. Comparison between the Technical Design and Manufacturing Characteristics of some Product Types

After laying down the representative type spectrum, an allocation process is carried out, in which the planned production programme is allocated to the types of the representative spectrum. Specific allocation criteria are to be established from case to case for this process. It is advantageous for this task to be assigned to a team of specialists drawn from the design and manufacturing staff. The next planning step is to code the information contained in the working documents with the aid of the classification system for the representative spectrum, and to store it on master tapes which constitute the technological data bank.

Depending on the problem statement, and by employing suitable evaluation programmes, the user receives the necessary demand profile from this data bank with little effort. In principle three planning stages are possible: 'rough' for preliminary studies, 'mean' for project tasks and 'fine' for detail studies. In the event of special planning tasks, supplementary data or information may be required. It should be mentioned that in a comparison process with the results of the offer side, the demand profile is already converted from a number of parts into machining hours.

The offer side proceeds in an analogous manner (Fig. 5.3, right-hand side). The plant and the equipment represent the starting position here. With the aid of the classification system applied already on the demand side, the plant and the equipment are coded accordingly. Finally, an offer is obtained in the form of the 'machine profile' which can be used as a comparable planning basis for the demand in the form of a 'machining profile'.

A direct comparison can now be made between the demand and offer at every planning stage, and from this results the necessary measures and consequences in respect of future investments and the personnel and corporate structure may be derived.

A major advantage of this methodical procedure is, that several variants can be prepared at the same time with a considerable reduction in effort in the case of extensive planning tasks.

5.2 Determination of the Representative Types Spectrum of Products

With a view to reducing the amount of effort required for the execution of technical investment planning, it has often been endeavoured to select a so-called 'representative types spectrum' from the area under investigation. An essential factor thereby is that the accuracy of the conclusions lies within the limits of the specified deviation areas. As the information concerning the representative types spectrum has only to be kept under constant control in this case, the effort required is kept within reasonable limits as opposed to a comprehensive consideration.

The deviation of the planned to the effective sales constitutes the starting point for determining the size of the representative types spectrum. The extent of the deviation is dependent on the realization probability of the planned quantity and characteristics of the types, which thus represent the parameter for determining the size of the representative types spectrum.

An example drawn from the pumps department of an engineering factory will be used to illustrate this statement. The starting point is provided by way of a comparison between the planned and effective total number of hours required to manufacture 15 types of pumps in a manufacturing area (Fig. 5.5). The scale diagram clearly shows the deviations, which can vary according to the type of products and market conditions and be caused by a number of factors. The resulting average deviation represents an important determining factor for the representative types spectrum. Starting from the principle of a planned yearly programme, the deviation of the total number of hours — on the basis of the complete consideration of the types spectrum compared with the total number of hours on the basis of the representative types spectrum — should not be larger than those of the external market influences.

This assumption was then investigated with the aid of an assumed yearly programme. As a first step all the respective product types were ascertained and coded with the help of the classification system mentioned in Chapter 4. This complete consideration showed a total of 138494 hours, which was taken as the reference value 100%. Fig. 5.6 A shows the 15 pump types and the corresponding total number of manufacturing hours for each pump type figuratively and graphically. From Fig. 5.6 B, can be seen the hourly related parts and machining profiles for the pump types.

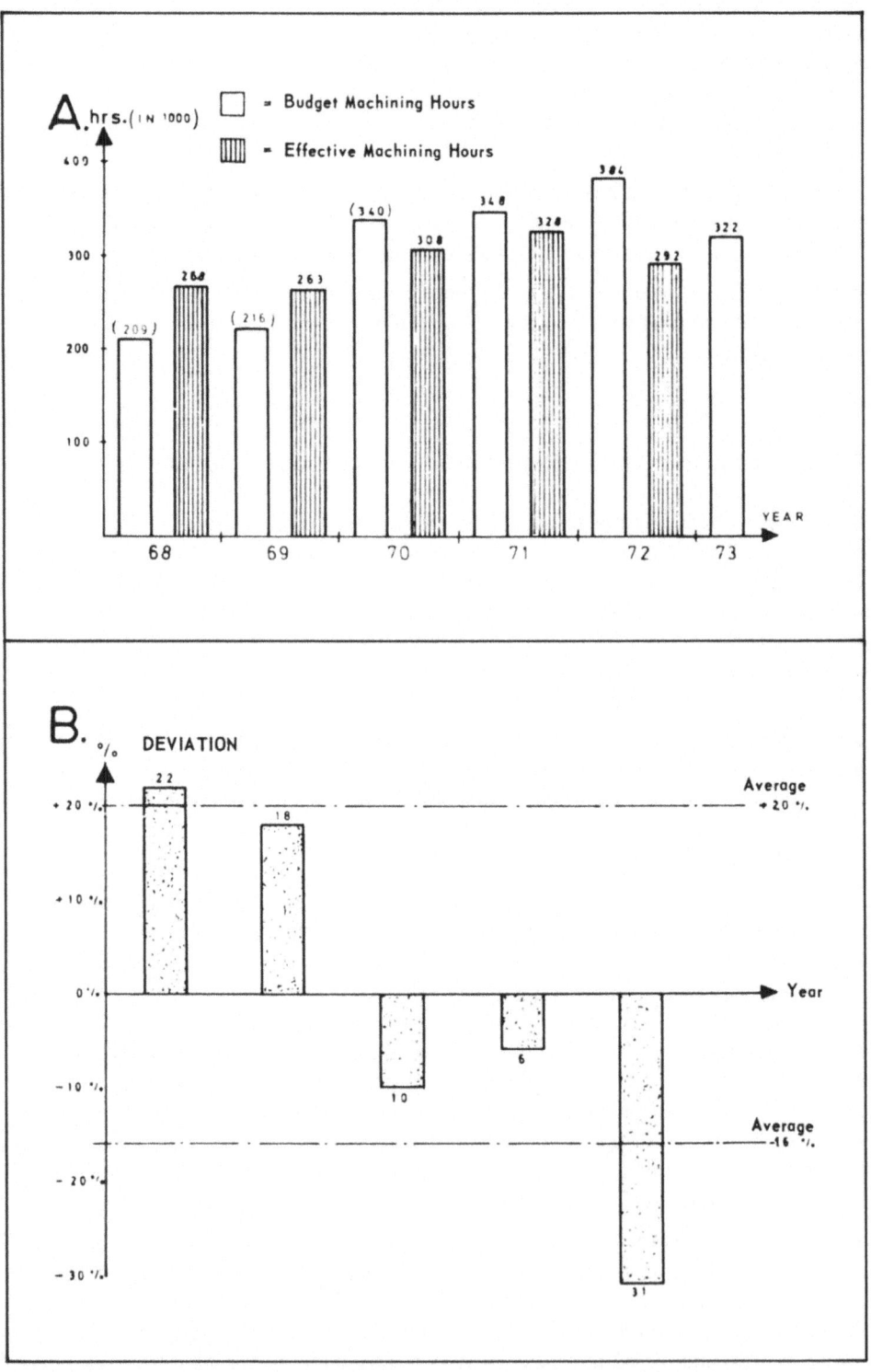

Fig. 5.5. Planned and Effective Total Number of Hours for the Production Programme of a Pump Area (A) Hours, (B) Deviation in Percent

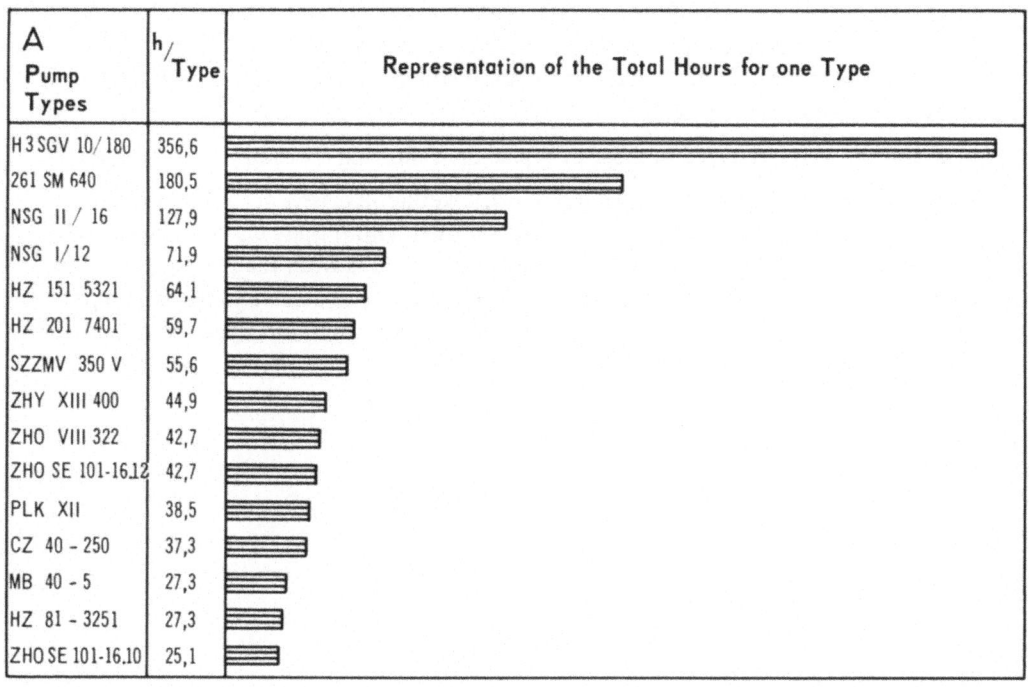

Table A – Type Programme and Total Hours

A Pump Types	h/Type	Representation of the Total Hours for one Type
H 3 SGV 10/180	356,6	
261 SM 640	180,5	
NSG II / 16	127,9	
NSG I/12	71,9	
HZ 151 5321	64,1	
HZ 201 7401	59,7	
SZZMV 350 V	55,6	
ZHY XIII 400	44,9	
ZHO VIII 322	42,7	
ZHO SE 101-16.12	42,7	
PLK XII	38,5	
CZ 40 - 250	37,3	
MB 40 - 5	27,3	
HZ 81 - 3251	27,3	
ZHO SE 101-16.10	25,1	

Table B

B Pump Types	Part Classes 0 Rotary Parts without Deviation	1 Rotary Parts with Deviation, General	2 Rotary Parts with Deviation, Casing, Housing	3 Non-Rotary Parts General	4 Non-Rotary Parts Casing, Housing	5 Forging, Pressing General	6 Forging, Pressing Casing, Housing	7 Casting, Sintering General	8 Casting, Sintering Casing, Housing	9 Assembly Groups	Machining Classes 0 Rotary Machining	1 Facing	2 Drilling, Boring	3 Manual Work General	4 Surface Treatment	5 Inspection, Checking	6 Heat Treatment	7 Non-Cutting Processing	8 Metal Joining	9 Original Forming Casting, Sintering
H 3 SGV 10/180	145,8	104,0	103,0	1,3	2,5						190,5	15,5	43,0	105,8					1,8	
261 SM 640	44,0	3,5	95,2	2,7	25,1						63,8	6,4	59,8	50,5						
NSG II / 16	52,2	56,7	19,0								74,7	6,9	6,9	99,4						
NSG I / 12	29,9	27,5	14,5								52,4	7,5	5,3	6,7						
HZ 151 5321	20,7	3,8	35,2	0,6	3,8						26,4	8,0	9,8	19,1					0,8	
HZ 201 7401	22,3	6,0	30,0	1,4							37,8	5,3	13,5	3,1						
SZZMV 350 V	14,2	5,1	25,2	10,8							22,1	3,7	12,5	12,3						
ZHY XIII 400	16,5	7,8	19,5	0,8	0,3						29,8	2,8	7,5	4,7					0,1	
ZHO VIII 322	13,0	6,2	22,9	0,6							25,4	0,6	6,3	10,2						
ZHO SE 101 - 16.12	11,3	12,4	18,3	0,7							28,9	1,1	7,4	4,9	0,1				0,3	
PLK XII	11,6	2,0	24,4	0,5							20,5	0,5	7,7	9,7	0,1					
CZ 40 - 250	8,5	11,4	17,4								21,6	2,0	4,7	8,7	0,1				0,2	
MB 40 - 5	8,5	7,9	10,9								17,6	2,6	3,0	4,1						
HZ 81 - 3251	12,8	2,6	11,0	0,9							16,2	2,7	6,4	2,0						
ZHO SE 101 - 16.10	6,6	9,9	8,2	0,4							15,6	1,3	3,9	4,3						

Fig. 5.6. (A) Type Programme and Total Hours – (B) Part and Machining Profile in Hours

5.2.1 Number of Representative Product Types

In order to ascertain the effects of the deviations — as compared with the overall consideration — the number of product types was gradually reduced on the basis of the mentioned investigation results.

In operation with a reduced number of types, the following procedure was observed:

To begin with, the various types were grouped together in classes according to their degree of similarity. Certain types were then omitted within these classes. Within each class the annual production number for the types excluded from the investigation were then allocated to so-called 'representative types'. In order to be able to determine possible errors resulting from the allocation process, an allocation into upper, middle and lower areas was carried out for each variant (Fig. 5.7).

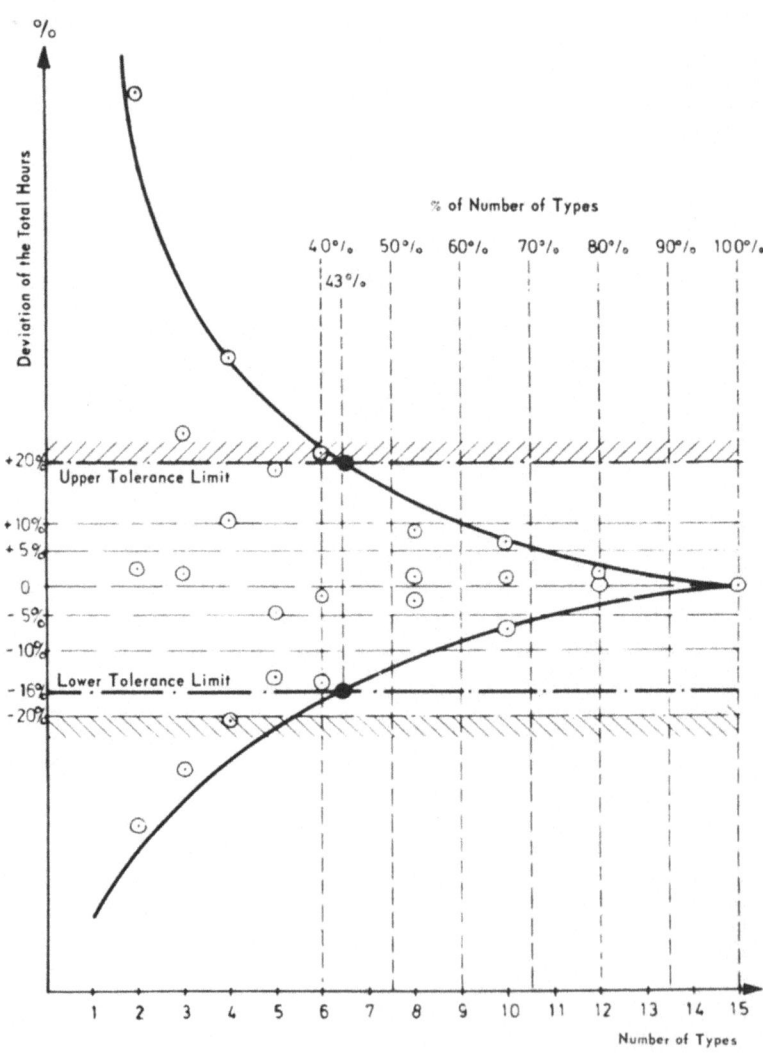

Fig. 5.7. Deviation of the Tital Hours in Relation of Number of Types

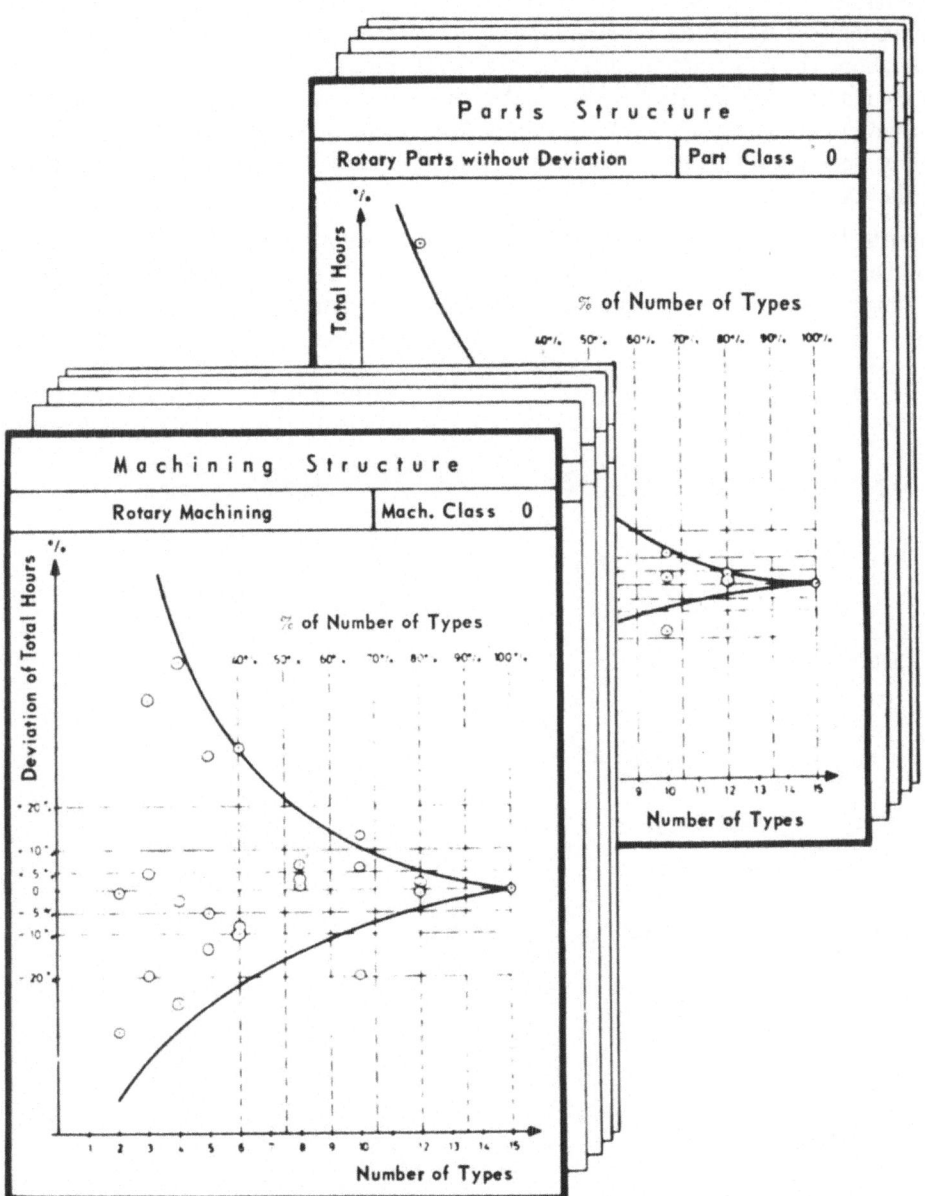

Fig. 5.8. Deviation of the Total Hours for the Machining and Parts Structure in Relation to the Number of Types

The percentual specified number of types represents the types spectrum laid down for the investigation which corresponds to a reduced number of types. The enveloping curve connects the largest deviations and thus represents the limiting values which also include the maximum allocation error. In planning the limits aimed for are provided by the average deviation between the planned and effective market quantity (Fig. 5.5).

The results, in this particular case (+ 20% and − 16%), are recorded in the graph which shows the deviation of the representative types spectrum in relation of the overall consideration. The minimum number of types, which ensures the result is maintained within the required and specified limits, can be determined from the intersection points of the enveloping curve and the straight lines. With this pump example, it is shown that it can be operated with 43% of the total number of types without impairing the accuracy of the statement.

Nevertheless, in the case of investment and layout planning knowledge of the total number of hours and particularly the respective allotment to individual part and machining classes is of major importance. To obtain such data, it is necessary to proceed

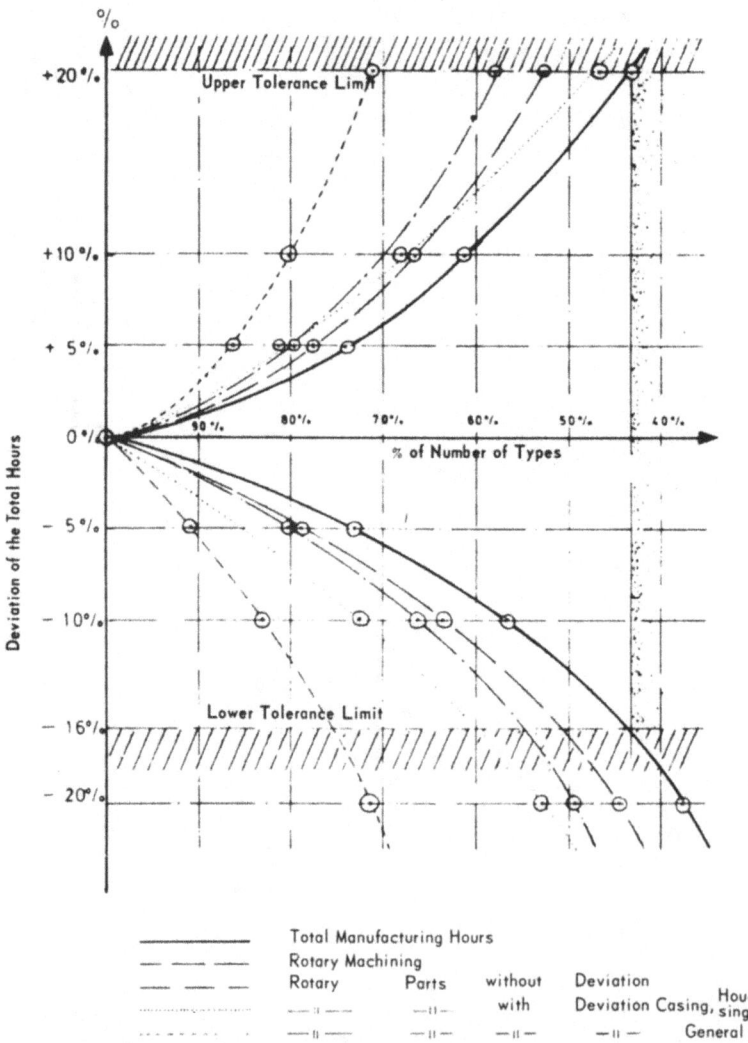

Fig. 5.9. Summary of the Structure Types and Tolerance Limits for the Deviation of the Total Hours

64

in an analogous way to the total number of hours. The result is shown schematically in Fig. 5.8. The number of types is again shown in relation to the deviation from the total number of hours. Fig. 5.9 finally presents a summarized representation which clearly shows that, for a specific tolerance field concerning the accuracy of statement, rough planning as opposed to fine planning may be carried out with a smaller number of types. The various scatter areas in the individual machining and part classes neutralize each other in the overall consideration, which also corresponds completely with the statistical view.

It results from the previous considerations concerning the representative types spectrum, that it can be worked with a relatively small representative type spectrum in the case of overall structure investigations. The number of types thereby is dependent on the deviation range of the external market influences. Detail studies for individual processes, however, call for a similar deviation margin of a larger number of types.

5.2.2 Type Mixture

The deviations from the number of pieces laid down for the individual types in the production programme, which may occur as a result of changes in the market situation, constitute a further important influential factor in the planning of manufacturing systems. In opposition to the number of types, which is based on the summarized total number of hours required, and by which the possible deviations may be ascertained with the aid of favourable and unfavourable market forecasts, the planning or the trend pattern for the individual types is even more difficult and associated with corresponding uncertainties.

No.	PUMP TYPES	Reference		Variant 1		Variant 2		Variant 3		Variant 4		Variant 5	
		Piece/y	h	Piece/y	h	Piece/y	h	Piece/y	h	Piece/y	h	Piece/y	h
1	H3SGV 10 180	38	13 550	15	5 349	5	1 783	10	3 566	55	19 613	80	28 528
2	261 SM 640	30	5 415	40	7 220	65	11 732	85	15 342	50	9 025	15	2 707
3	NSG II / 16	50	6 395	90	11 511	115	14 708	150	19 185	25	3 198	80	10 232
4	NSG I / 12	120	8 628	50	3 595	25	1 798	60	4 314	30	2 157	150	10 785
5	HZ 151 5321	295	18 809	335	21 474	400	25 640	100	6 410	200	12 820	230	14 743
6	HZ 201 7401	75	4 418	85	5 075	100	5 970	250	14 925	400	23 880	50	2 985
7	SZZMV 350 V	60	3 336	20	1 112	15	834	85	4 726	75	4 170	25	1 390
8	ZHY XIII 400	60	2 694	100	4 490	140	1 296	75	3 368	25	1 122	15	674
9	ZHO VIII 322	175	7 473	205	8 754	58	2 477	25	1 068	310	13 237	100	4 270
10	ZHOSE 101–16.12	315	13 450	400	17 080	500	21 350	600	25 620	200	8 540	300	12 810
11	PLK XII	360	13 860	250	9 625	100	3 850	50	1 925	230	8 855	150	5 775
12	CZ 40 – 250	300	11 190	600	22 380	750	27 975	500	18 650	200	7 460	50	1 865
13	MB 40 – 5	500	13 650	50	1 365	25	683	400	10 920	290	7 917	700	19 110
14	HZ 81 – 3251	200	5 460	350	9 555	450	12 285	170	4 641	300	8 190	230	6 279
15	ZHOSE 101–16.10	405	10 166	435	10 918	270	6 777	375	9 413	315	7 907	540	13 554
	$\sum$	2983	138 494	3025	139 503	2918	139 158	2935	144 073	2705	138 091	2715	135 707

Fig. 5.10. References for the Model Test of Type Mixture

Reference	Variant 1	Variant 2	Variant 3	Variant 4	Variant 5	Machining
78'942 100 %	78'819 99,84 %	76'646 97,1 %	83'562 105,8 %	78'021 89,83 %	78'931 89,75 %	Rotary Machining
8'864 100 %	9'207 103,9 %	9'493 107,1 %	8'644 97,52 %	8'700 98,15 %	8'821 99,52 %	Facing
21'744 100 %	22'460 103,29 %	23'100 106,24 %	24'493 112,6 %	25'133 115,59 %	19'271 88,7 %	Drilling, Boring
28'944 100 %	29'017 100,25 %	29'919 103,37 %	27'374 94,6 %	26'237 90,65 %	27'684 95,65 %	Manual Work, General
138'494 100 %	139'503 100 %	139'158 100 %	144'073 100 %	138'091 100 %	135'707 100 %	Total Machining

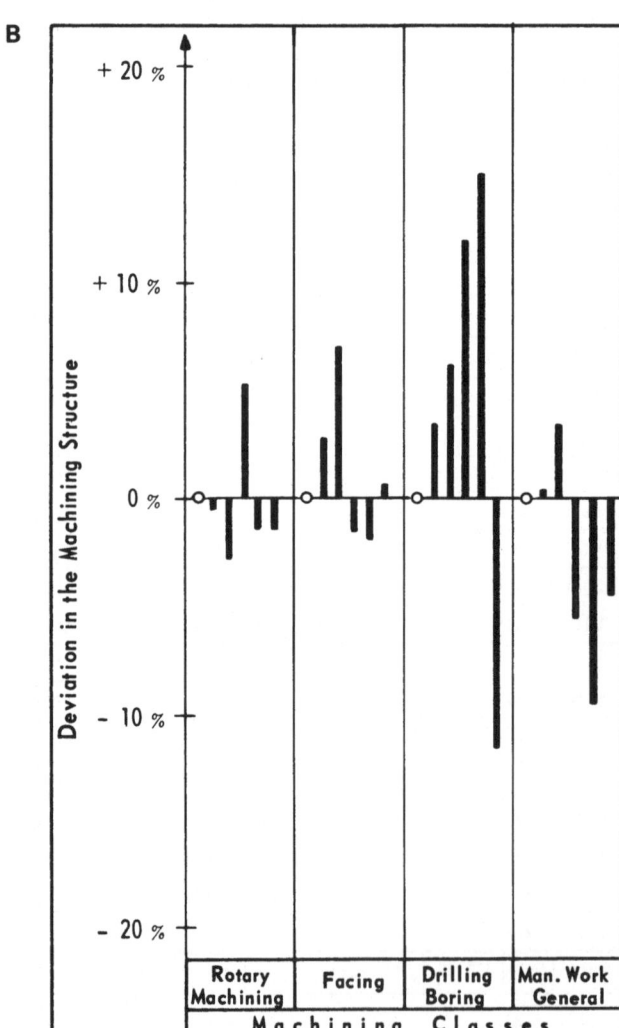

Fig. 5.11. Effect of Various Type Mixtures on the Machining Structure with a Constant Total Number of Hours.
(A) Absolute Values,
(B) Deviations in Percent.

With a view to elaborating this concept even further, the following shows a model test based on 5 variations of the type mixture. The marginal condition is that all the variants exhibit the same total number of hours such as the reference value, which was used at the first test (Section 5.2.1) to determine the deviations in the case of the number of types. In the type mixture simulation, the annual number of 15 pump types, based on the reference value, was varied in accordance with the practical possibilities. In Fig. 5.10, the annual number of pieces and the corresponding number of manufacturing hours are shown for each variant. Several characteristic machining classes have been evaluated on an hourly basis from the machining structure of the product programme. The absolute values as well as the deviations to the reference value, which is always set at 100%, can be seen in Table A of Fig. 5.11. Fig. 5.11 B, however, shows the deviation within the various machining classes in a graphical form. During the model test, maximum deviations of approximately ± 17% were noted.

As a result of this investigation, the following planning recommendations can be made:

— The information appertaining to the demand and offer sides should be evaluated according to a procedure which can be divided into rough, mean and fine stages.
— Dependent of the required planning objective, the corresponding stage should be selected for the problem solution.
— In general, extensive investigations and/or with those where a large number of variants are required should be carried out with the aid of a representative type spectrum. If empirical values are not available, the representative number is to be determined preferably by means of random checks in an analogous manner of the model example.
— In establishing the representative number of types, it should be considered whether the investigation is related to the overall structure or individual processes.

5.3 Layout Planning

One of the most important aspects of a functional layout system is the realization of the best possible loading for every individual equipment. With GT-flow lines or cells, however, due consideration is given primarily to a reduction in the total throughput time. Although a loading optimum is aimed here as well, this condition only constitutes the second priority. The two conditions can only be met simultaneously in the ideal case. Using this objective as a basis, the task of layout planning within the framework of GT is to associate a specified machining group with the optimum rationalization stage of the GT-manufacturing system (line, cell or centre).

The fundamental position is constituted by the machining profile resulting from the investment planning on the basis of the product and production programme. Based on that, the parts spectrum is firstly subdivided into similar machining groups and the annual amount of the necessary hours is then calculated. The degree of similarity of the machining groups is then used to determine the corresponding stage of the GT-layout forms. This is effected in an iterative procedure, by which an attempt is made to obtain a higher stage of GT-layout by raising the degree of similarity. This process is assisted by two working aids, namely through the design family for the purpose of increasing the degree of similarity in respect of design — as dealt with in Chapter 6 — and the machining family for raising the degree of similarity in respect of machining in the process planning phase, which will be described in Chapter 7.

The selection of machines is made after the optimum stage of the GT-layout has been determined. In the case of new planning, the selection is based on the machine market, in the case of rearrangement on the existing machines. In practice these two forms will be mixed.

Generally speaking, in practice consideration is firstly given to the layout planning phase, which is based on less detailed investigations than is the case with the design and machining families.

The indirect results of this planning of layout forms will influence the design and machining families. The similarity data being the direct result of this investigations will again affect the layout forms, and so the entire process is given a rolling character. Fig. 5.12 shows the three stages of this circle of problems and their interrelationships.

5.3.1 Determining the Machining Groups

In connection with the product and production programme, which constitutes the basis for the investment planning, and through which the total hour demand and the type structure is obtained, several loading variants are investigated for the individual machining groups and indicated in a loading plan. Fig. 5.13 provides a schematic representation of the evaluation programme for a rough determination of the machining groups. It is based on the technological data bank, in which the planning data for the parts spectrum of the representative product types are stored. This information can be called up for investigations and evaluated according to the similarity criteria input.

One important aspect that should be mentioned is, apart from the classification characteristics for describing the similarity, that the identification of the parts by means of an article number and the cost centre of the individual operations appear in the loading plan. In the case of rearrangement planning, the rationalization task consists of relocating a specified manufacturing area, an action which entails selecting a parts spectrum out of an overall one. This means that individual operations can be executed far off in other areas. The establishment of the limits of investigation will always be one of the major problems. The two procedure stages — described in Figs. 5.12 and 5.13 — offer an aid with which we can approach the ideal conception of a minimum total throughput time for the parts with maximum loading of the individual equipment. From the overall economic standpoint, the selected degree of loading should be derived within the framework of the respective manufacturing system, whereby a differentiation should be made between primary and secondary machines.

Apart form the individual operation times, the sequence of working operations, which are important for the laying out of GT-lines, is also shown in the loading plan. To sum up, the establishment of machining groups represents an iterative procedure between the degree of similarity of the machining group and the machine loading, coupled with the resultant design and machining families which enables the degree of similarity to be increased.

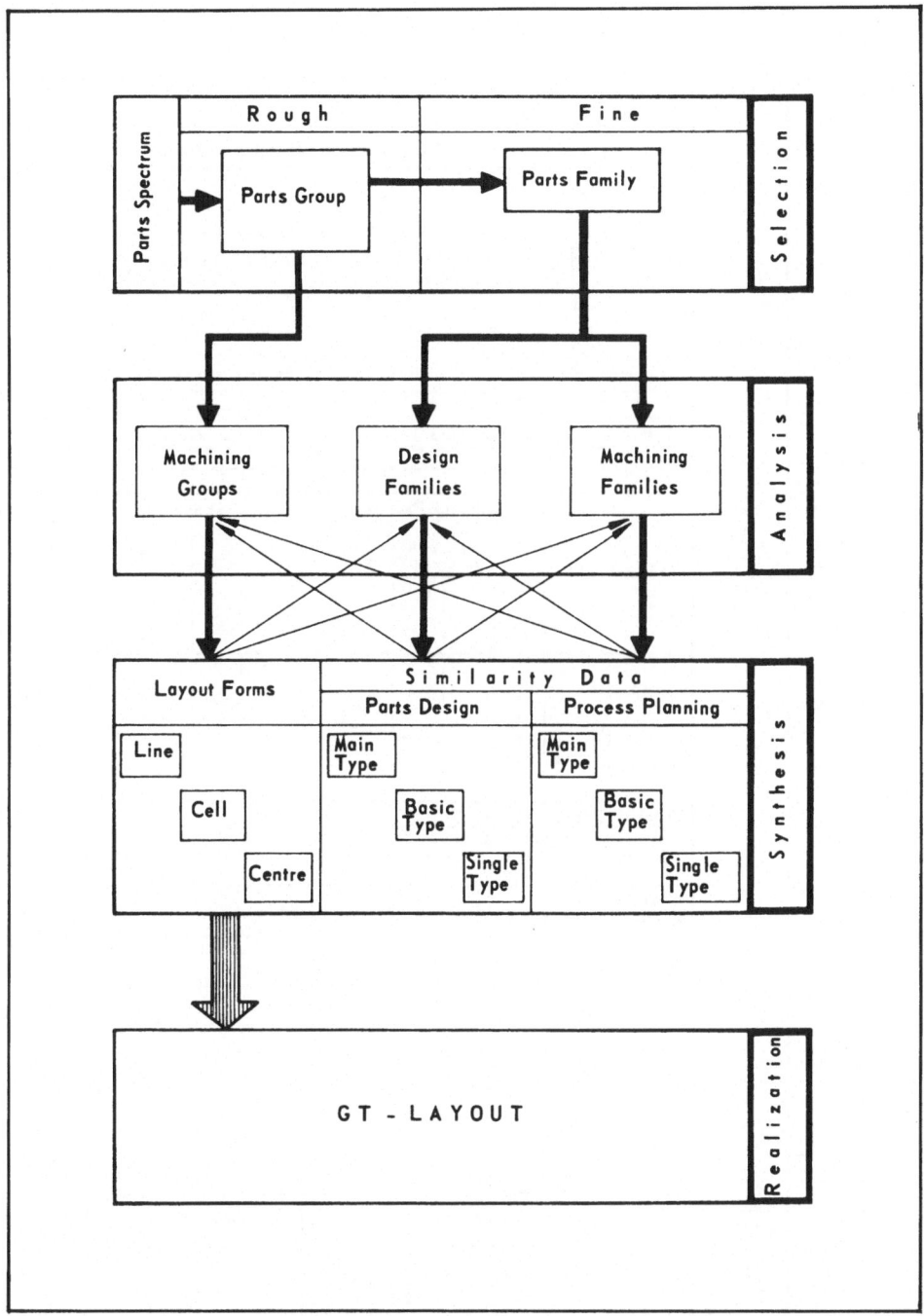

Fig. 5.12. Procedure Stages for the Layout Planning and Interrelationship with the Design and Machining Families

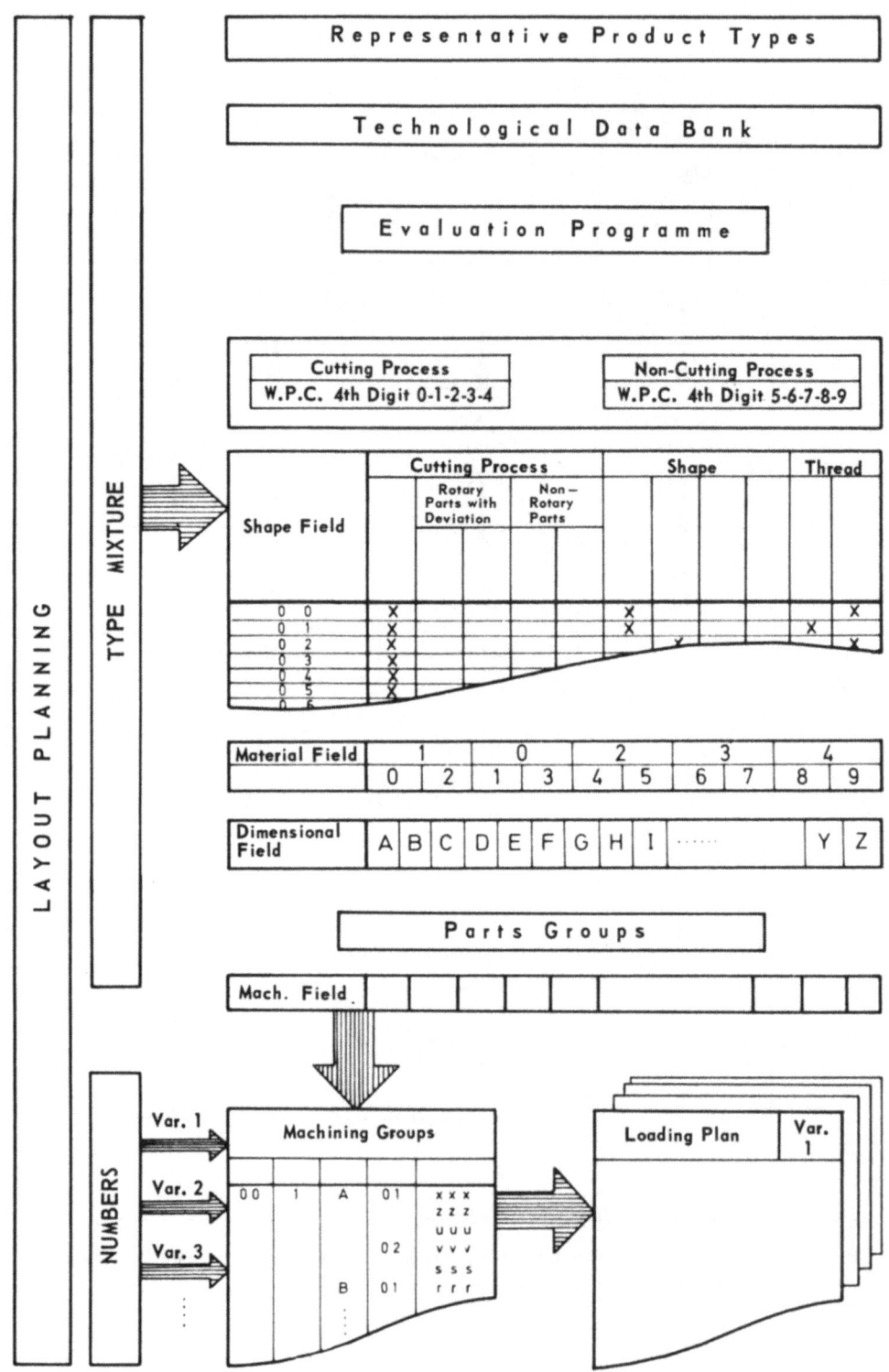

Fig. 5.13. Procedure Diagram for Determining the Parts and Machining Groups

5.3.2 Determining the GT-Layout Form

The next procedure phase entails determining the optimum layout form for the selected machining group. As already stated, a stepwise procedure is adopted, the pattern of which is schematically shown in Fig. 5.14. The loading plan for the selected machining groups and the necessary variants for obtaining an economic load are shown once more at the top of the mentioned figure. The resulting loading plan provides the starting basis for determining the GT-layout form. This is effected by means of the investigation of the sequence. If a high degree of conformity is attained, there is a chance for a GT-flow line. If, however, the preliminary conditions are not met, an attempt is made by means of standardization in the design and process planning to obtain a higher similarity degree before the second rationalization stage is investigated. The various stages are examined in an analogous manner. The component remaining after the fine analysis has been made is returned to the functional layout. Fig. 5.15 shows a section of a fine analysis of the working operation sequence for an investigated machining group. The process consists of a stepwise procedure. To begin with, the parts belonging to the machining group are subdivided into groups with the same working operations. These are given the designations A, B, C. After this, the parts of every one of these groups are investigated for similarity of the working operation sequence. Those parts with the same working operation sequence are designated A1, A2, etc., and constitute subgroups which are suitable for optimum flow line layout.

In the evaluation phase, the systematic procedure is very suitable for using a computer programme. Nevertheless, the loading of the quipment is one of the economic aspects that have to be considered before such a GT-flow line can be realized.

If the preliminary conditions for a GT-flow line have not been met, an analogous procedure may be used for the GT-cell. In such a case, the investigation is limited to the working operations only, because the limitations in respects of the working operation sequence have no influence on the cell.

In the case of a GT-centre, the analysis is only related to the individual working operations. Once the analysis has been completed, the parts lists and the loading plan are brought in line with the selected GT-layout form and then submitted to the decision-making authority.

5.3.3 Selection of Machines

After having cleared the problem of the machining groups, which were grouped together in the previous phase, the next operation is devoted to the selection of the respective machines. Consideration must be given thereby to the probable deviations by changes in the market within the framework of investment planning. This particularly applies to so-called 'bottleneck' machines.

The coded workshop drawings constitute the basis for determining the demand profile (Fig. 5.16). The form of coding employed is machine-neutral and does not consider the special aspects of the individual machines. That is sufficient for a rough investigation. In numerous cases, however, there is no conformity between the workpiece dimensions and the required working area. This means that the demand profile has to be modified for fine planning, particularly in the machine selection phase. This is directly dependent on the selected types of machine and therefore on the machining processes.

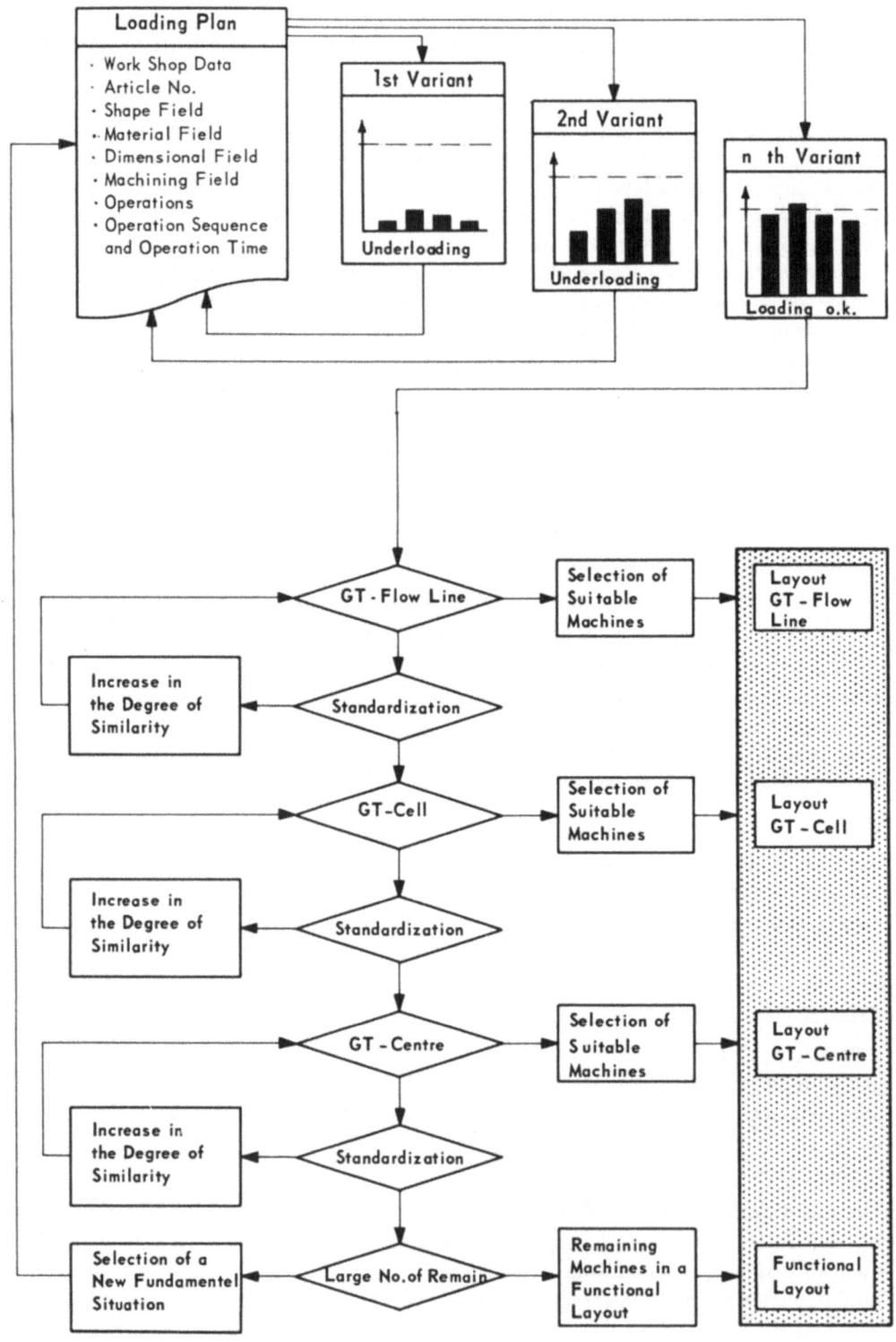

Fig. 5.14. Layout Planning Procedure for Determining the Optimum Layout Form

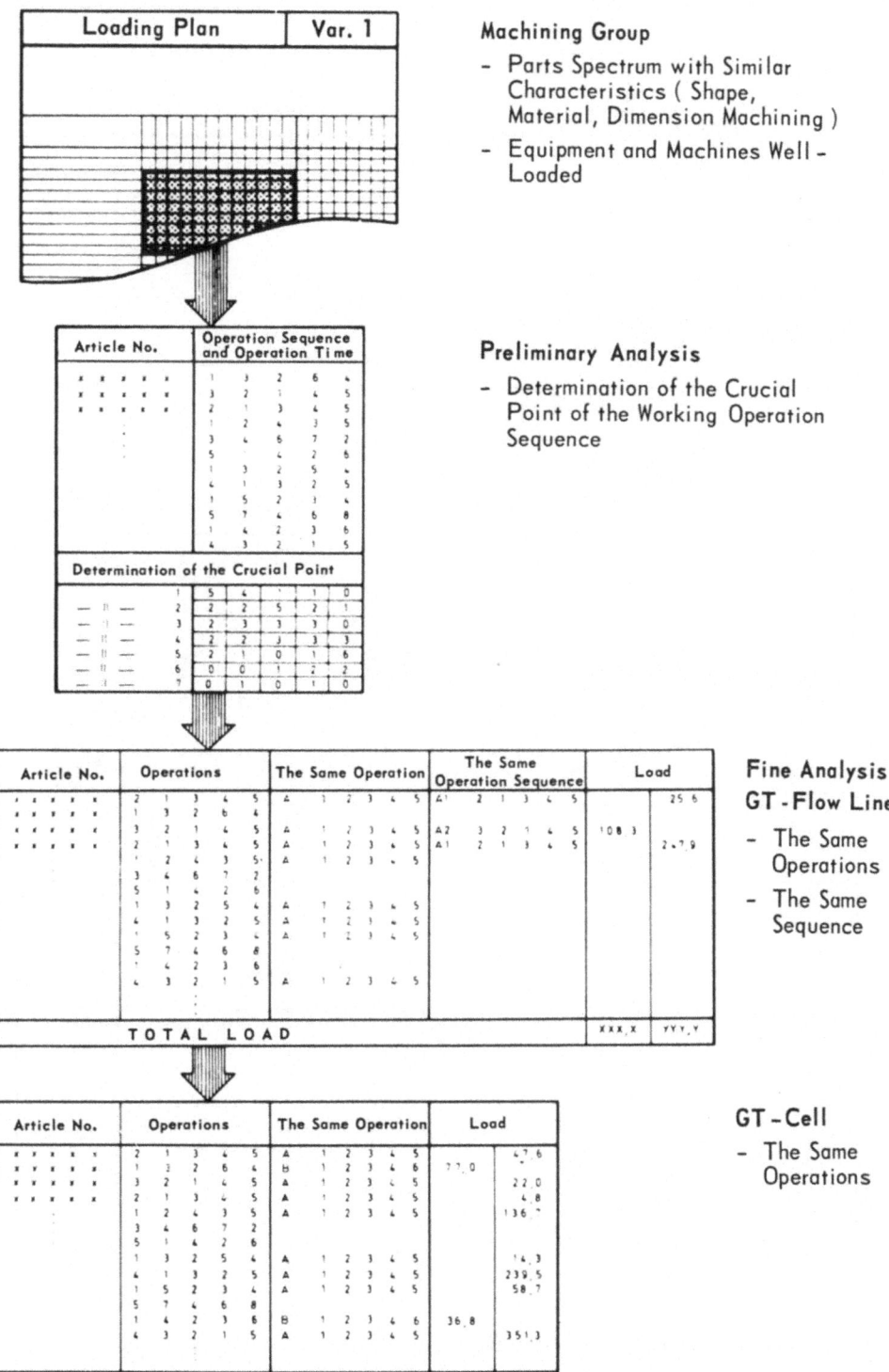

Fig. 5.15. Preliminary and Fine Analysis for Determining the GT-Manufacturing Forms

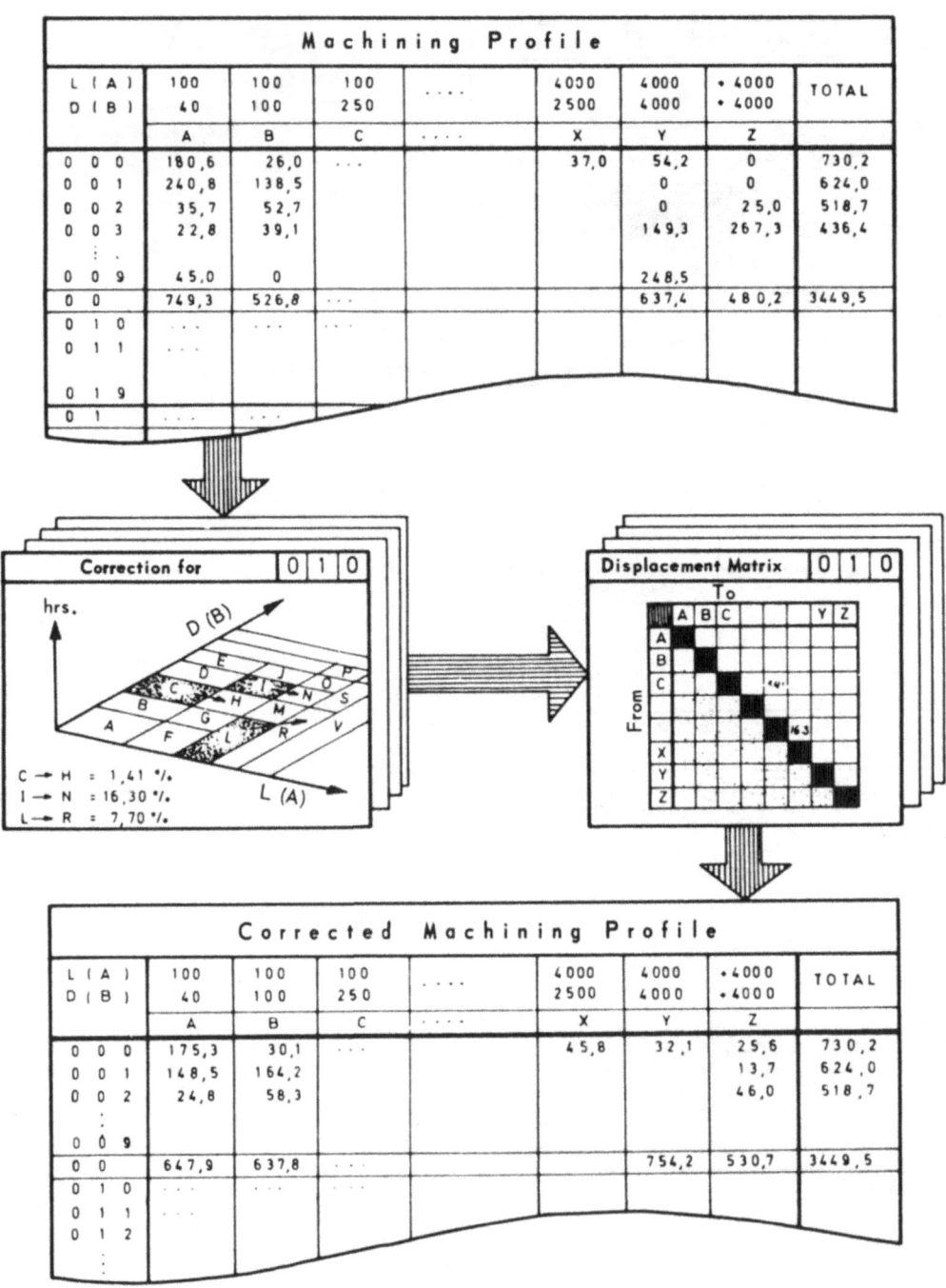

Fig. 5.16. Requirement Profile in Relation to the 26-figured Dimensional Classes

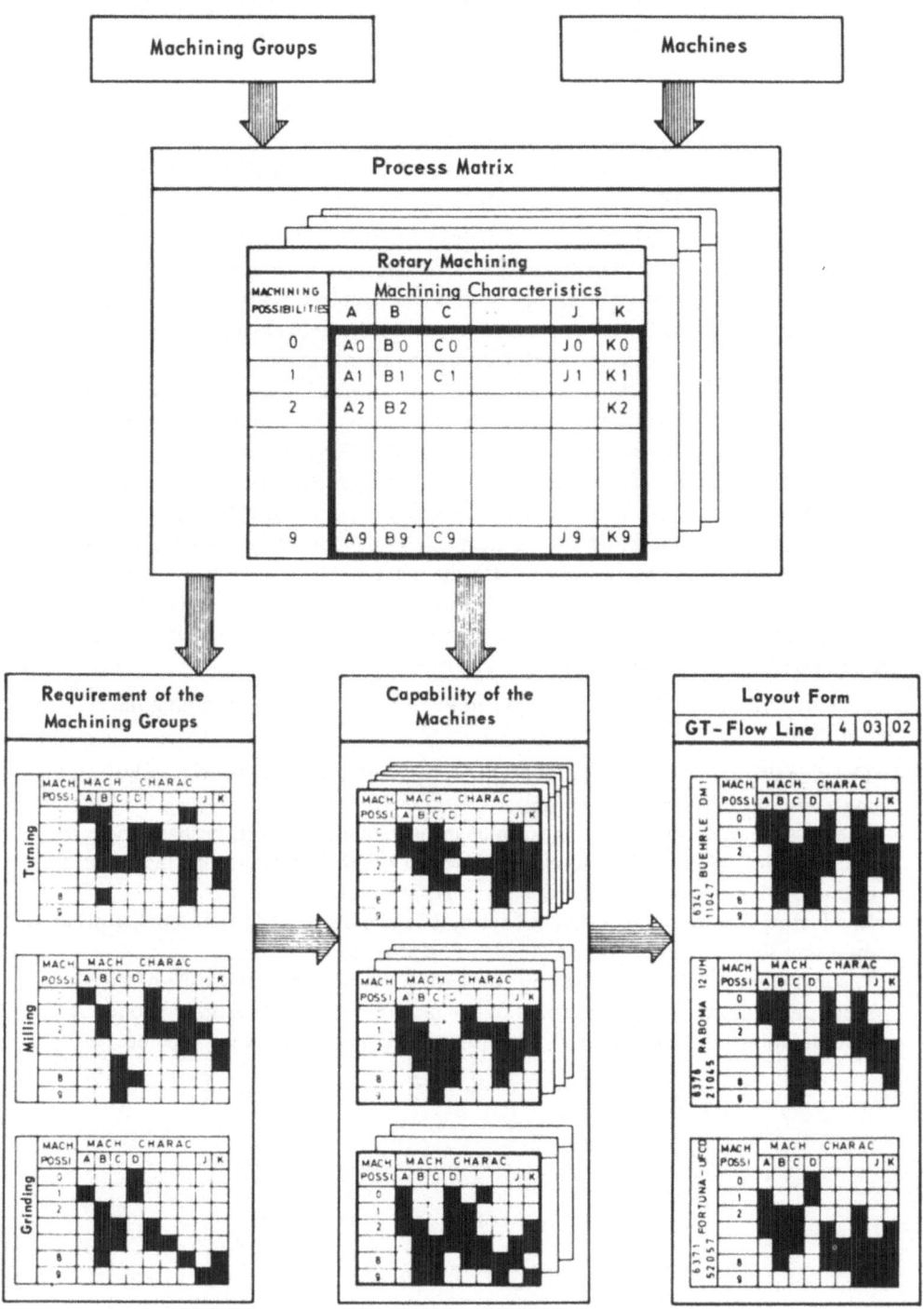

Fig. 5.17. Procedure for the Selection of the Equipment for a Certain GT-Manufacturing Form

The correction can be of a positive or negative character. Internal boring with a large overhang or drill units, which can be mounted on large workpieces during the assembly stage, may serve as an example.

In order to effect this correction with the computer as well, correction matrices, which are ascertained from the parts spectrum by means of a random check, are employed. This enables the demand profile for the investigated machining group to be corrected according to the proposed types of machine.

As already stated, the machining requirements of the workpieces can only be roughly defined with the aid of the classification system described in Chapter 4, and this is not sufficient for the selection of machines. In this case, detailed information concerning both the parts as well as the machines and related to the individual processes is necessary. It is possible to execute this partial task in a rational manner with the aid of operation matrices [33]. The matrices, which have to be ascertained for every process, e.g. rotary machining, facing etc. are obtained by associating the machining possibilities with the various specific process machining characteristics. The parts spectrum requirements and the capability of the machines in question can then be determined on the basis of these matrices. Such forecasts can be directly compared with each other (Fig. 5.17). That machine which meets the machining requirements and provides the optimum degree of coverage, is selected for the manufacturing system. However, it must be admitted that other economical factors may also have to be considered. In the event that the machining requirements are not covered by any machine, the data will represent the desired criteria for an investment.

To sum up, the following basic rules have to be observed when determining the GT-layout form:

— The planning effort can be reduced without impairing the conclusiveness by employing a representative types spectrum.
— The possible deviations resulting from changes in market conditions have to be accounted for in planning, because they provide us with important information in the determination of bottlenecks.
— In selecting the optimum GT-layout form, it is advisable to adopt a stepwise procedure that simplifies and reduces the planning procedure.
— The machines should be selected by comparing the requirement profile of the parts spectrum with the capability profile of the machine. Apart from other economic considerations, the machine with the optimal degree of coverage is selected.

6 Parts Design

Within the framework of parts design and the elaborating of drawings, the application of GT is particularly effective because it represents one of the most work-intensive activities within the product design field of mechanical engineering. Attempts were therefore made at a very early stage to rationalize this crucial point. For that purpose the following methods are generally known

— Standard parts,
— Recurring parts,
— Similarity types.

When these rationalization methods are considered, we find that they can lead to useful results individually, but do not provide a unified consideration or reciprocally limit the fields of application. Furthermore, as a result of the more or less intuitive procedure used in selecting the rationalization methods and compiling the working aids, large portions of the complete parts volume will inevitably not be considered in some cases.

On the basis of this situation, a solution was sought which would place the entire problem complex in a unified framework. Using a procedure as an aid, the basic idea therefore is to determine the method to be applied with the relevant working aids from the prognosis of the expected application frequency for a selected parts spectrum.

A special method is introduced to build up the similarity types. This provides for a step-by-step standardization process on the basis of defined similarity stages in the design and enables us to cover the largest possible percentage of the parts spectrum.

The term 'design families', which pre-groups the parts spectrum to be investigated in accordance with the characteristics of shape, material, function and dimensions, is based on the classification system for workpiece description and provides the starting point for our consideration.

6.1 Survey of the Rationalization Methods

We firstly characterize the various rationalization methods for the parts design. The GT influential factors and the aspects of the working aids, which, in the sense of the required overall consideration, are complimentary to each other, stand in the foreground.

6.1.1 Standard Parts

Apart from reducing the number of types and sorts in the technical field, the term 'standardization' infers the unification of sizes, shapes, materials, quality grades etc., of the material and workpieces. The result is called 'standard'. Besides the international standards, there are also national and firm-related standards. In the case of the group of problems under consideration, the firm-related standard parts lie in the foreground, because they constitute the link to the recurring parts and similarity types which are dealt with in detail in this investigation.

The common features of these standard parts are that all the characteristics are laid down and therefore allow no further degree of freedom in the parts design. Therefore it is the aim to work with standard components in the design of main assemblies or complete products — as is already extensively the case in the electronic field — with a view to avoiding work-intensive design detail work.

The rationalization aims of GT are also directed towards this ideal goal, in that an attempt is made to increase gradually the degree of similarity in the parts spectrum. These endeavours are also supported by the design guidelines, for they group the various recommendations for the elaboration of single parts together. Apart from the general rules and those governing the component and element shapes, such recommendations are principally matched to the machine-technical requirements of the various manufacturing fields. With the increasing degree of automation in manufacture, they will become more important and have to be considered in many cases already in the conceiving and outlining stages.

6.1.2 Recurring Parts

A further recognized method for the reduction of parts variety and production of drawings are the recurring parts. By means of this rationalization measure, an attempt is made to obtain a selection of representative drawings from a similar parts spectrum and to use it as information, e.g. in the form of a so-called recurring parts catalogue, with a view to reducing the parts variety. Apart from the general similarity criteria, and depending on the complexity of the parts, further specific design data can be provided in the catalogue, which can serve as an objective-orientated selection of the drawings under consideration. From the manufacturing standpoint, it is expedient to supplement the selected drawings by a classification of the average yearly consumption of the recurring parts. Fig. 6.1 represents an extract from such a catalogue. The index is provided with symbols to simplify the search work. The heading contains the general information, whilst the lower part gives the individual data for the recurring part and is designed in such a way that the computer output can be used directly. This form of rationalization is mentioned in the literary references [34] and has already been successfully introduced in several firms.

6.1.3 Similarity Types

Similarity types are placed between the recurring parts and the standard parts. In opposition to the recurring parts catalogue, which takes over the individual drawings in an unchanged condition, a step-by-step standardization, in accordance with the existing parts frequency, is aimed for in the build-up of similarity types. In an analogous manner to standardization, the parts frequency also serves as a basis for the selection of rationalization. A frequency analysis according to the main and secondary shape is sufficient in many cases for the decision preparation, this being effected with the help of the workpiece code. If the analysis should show no pronounced crucial areas in respect of the similarity structure, then the recurring parts catalogue may be selected advantageously because of the reduced effort. If, on the other hand, the parts spectrum to be investigated shows pronounced crucial areas, then the effort involved in the standardization process for the purpose of obtaining similarity types is worthwhile.

The proposals made in the literary references [3, 22, 34, 35, 36] for the standardization of parts in one-off and small batch production are mainly based on a general work-

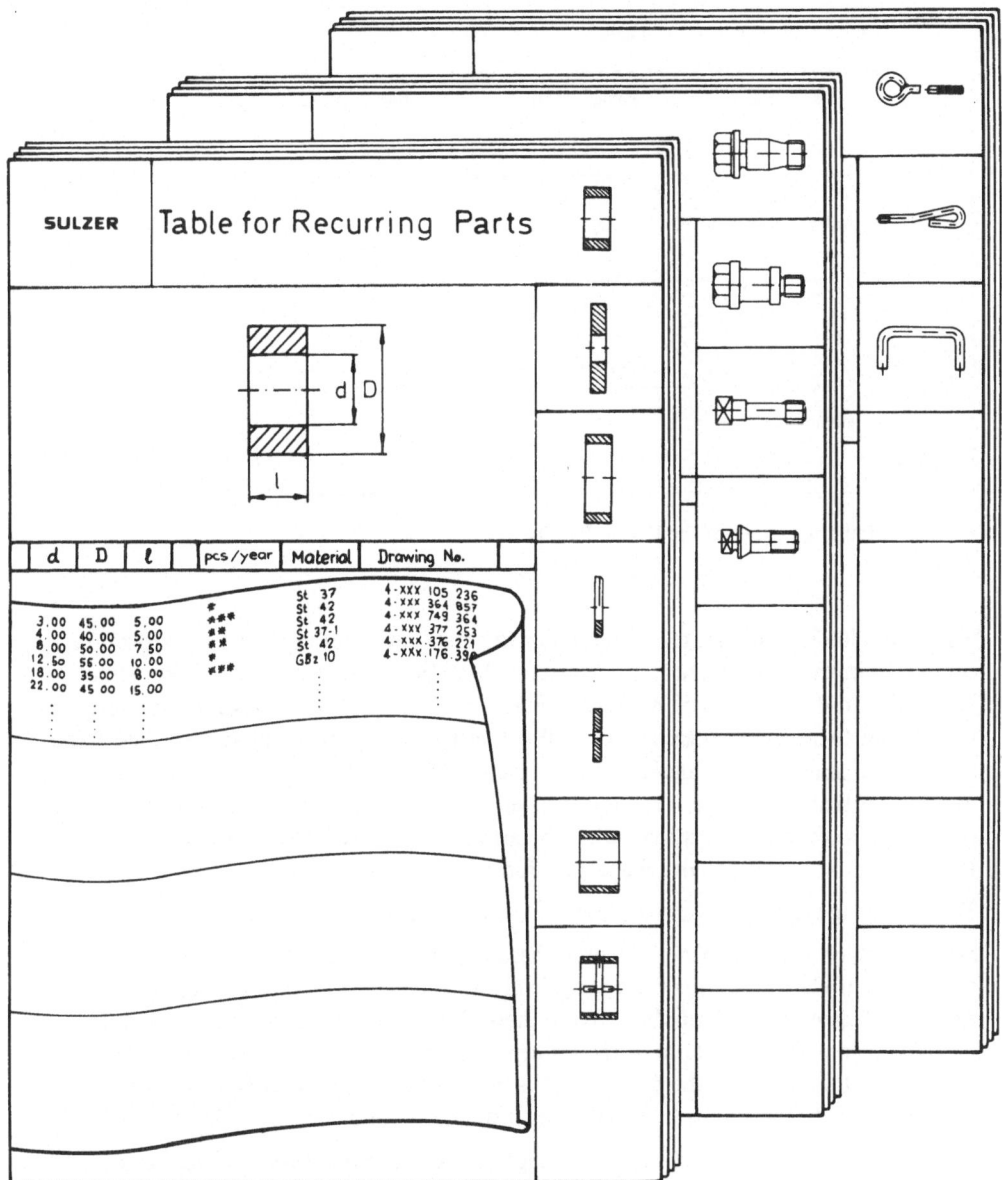

Fig. 6.1. Recurring Parts Catalogue

piece classification and a subsequent intuitive evaluation. Other investigations were carried out with the aid of a more refined workpiece classification, but these require a great deal of effort in the preparation stage if the specific standardization problems of the individual design families are to be completely covered. The solution proposed in this work is based on a relatively rough, but comprehensive workpiece description classification system, and is described in more detail in Chapter 4.

Instead of an intuitive procedure, a systematic one has been developed which, with the aid of defined similarity stages, enables the design families to be investigated and ana-

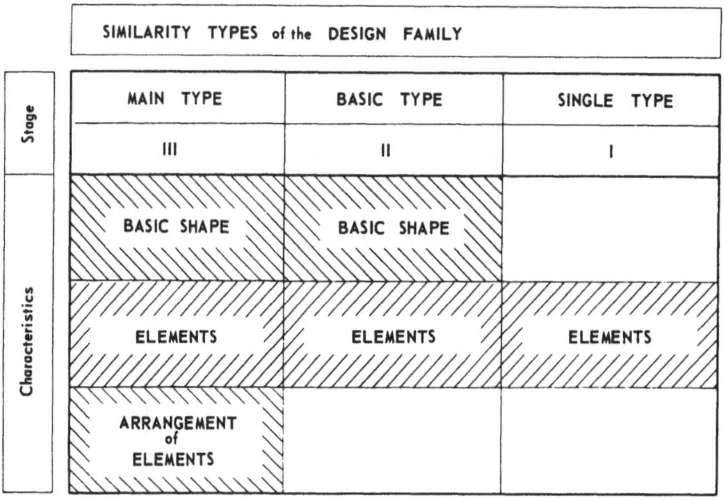

Fig. 6.2. Similarity Types of the Design Family

lysed completely, and creates the basis for the elaboration of unified standardization proposals. In this case, three similarity stages have been selected, which differ from each other in their degree of similarity (Fig. 6.2). The extensive investigations have shown that these three stages can be defined with the help of the characteristics 'basic shape', 'elements' and 'arrangement of elements'. The basic shape applies to the abstract external and internal shape. The elements represent the individual functions on the work-piece, e.g. bores, grooves, thread, etc. The arrangement of elements describes the allocation of parts according to the shape position within the basic shape.

The following gives a brief description of the three stages.

Stage III (main type) provides all three characteristics — basic shape, elements and arrangement of elements. This third stage leads to so-called main types which only have a limited degree of freedom. Apart from these three characteristics, further design data and dimensions — in the sense of recommendations — may be included. If we decide to standardize this data and the dimensions — in the light of the frequency analysis — then this already represents a changeover to standard parts.

Apart from the standardized elements, stage II (basic type) covers the basic shapes, which are selected with the aid of the similarity analysis, and standardized in the subsequent synthesis. The arrangement of elements — as the third characteristic — is not standardized by this stage; the analysis results show no semblance of similarity frequency in this respect. This standardization stage is known as the basic type, because it is based on a standardized basic shape.

Stage I (single type) is restricted to the standardized elements within the framework of the allocated design family. The basic shape and the arrangement of elements remain free because the frequency analysis shows no similarity characteristic in the basic shape and arrangement of elements. This class is known as the single type within the framework of design families because — with the exeption of the standardized elements — there is a great deal of tolerance in respect of the basic shape and the arrangement of elements.

6.1.4 Comparison

Fig. 6.3 shows the degree of standardization and the corresponding degree of freedom in the parts design for the three rationalization methods. The allocation to a particular stage is carried out with the aid of a frequency analysis, which also serves as a working aid for forecasting the cases of application to be expected in the future.

The recurring parts constitute an exception, however, for the effort is required for classification and grouping and no modifications in the sense of standardization are necessary. From the degree of freedom standpoint, the recurring parts resemble the standard parts.

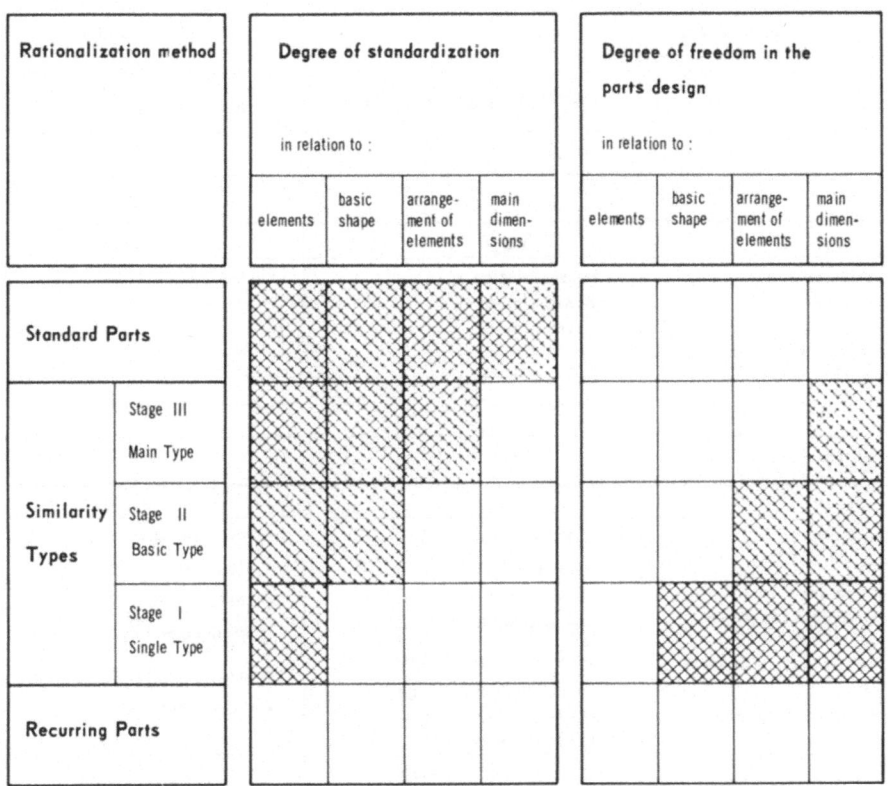

Fig. 6.3. Comparison of the Rationalization Methods

6.2 Procedure: Similarity Types of the Design Family

A procedure, which provides for the following phases of work, has been developed for the purpose of obtaining a rational approach (Fig. 6.4):

— Selection of the rationalization method,
— Analysis of the design family,
— Synthesis,
— Building-up the working aids.

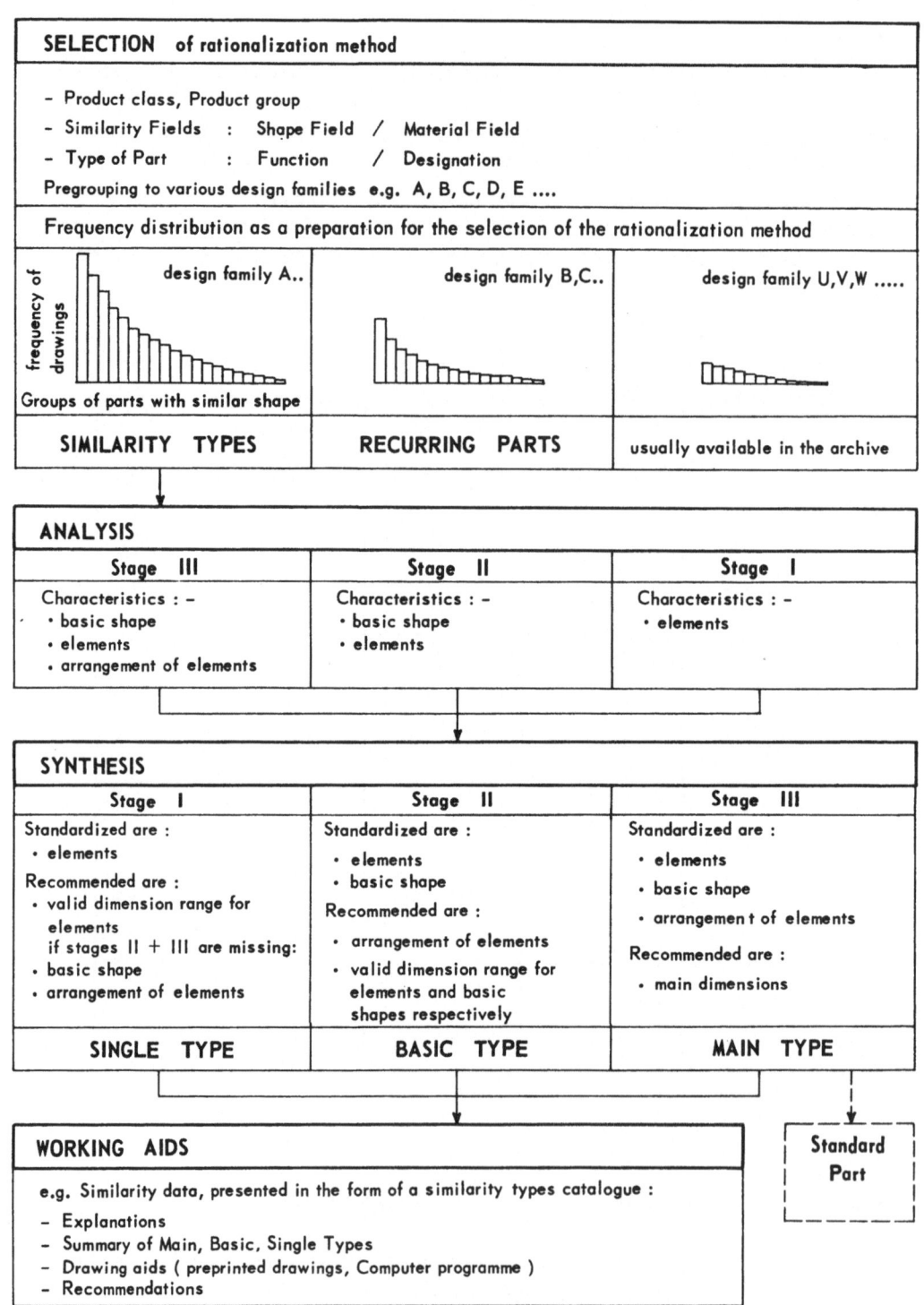

SELECTION of rationalization method

- Product class, Product group
- Similarity Fields : Shape Field / Material Field
- Type of Part : Function / Designation
Pregrouping to various design families e.g. A, B, C, D, E ....

Frequency distribution as a preparation for the selection of the rationalization method

| frequency of drawings | design family A.. | design family B,C.. | design family U,V,W ..... |

Groups of parts with similar shape

| SIMILARITY TYPES | RECURRING PARTS | usually available in the archive |

ANALYSIS

Stage III	Stage II	Stage I
Characteristics : –	Characteristics : –	Characteristics : –
• basic shape	• basic shape	• elements
• elements	• elements	
• arrangement of elements		

SYNTHESIS

Stage I	Stage II	Stage III
Standardized are :	Standardized are :	Standardized are :
• elements	• elements	• elements
Recommended are :	• basic shape	• basic shape
• valid dimension range for elements	Recommended are :	• arrangement of elements
if stages II + III are missing:	• arrangement of elements	Recommended are :
• basic shape	• valid dimension range for elements and basic shapes respectively	• main dimensions
• arrangement of elements		

| SINGLE TYPE | BASIC TYPE | MAIN TYPE |

WORKING AIDS

e.g. Similarity data, presented in the form of a similarity types catalogue :

- Explanations
- Summary of Main, Basic, Single Types
- Drawing aids (preprinted drawings, Computer programme)
- Recommendations

Standard Part

Fig. 6.4. Procedure Plan: Similarity Types of the Design Family

82

In this way, the defined similarity types can be derived and the individual steps developed as follows.

6.2.1 Selection of the Rationalization Method

The selection criteria for the creation of design families and the subsequent frequency investigation for determining the rationalization method to be selected are based on the product main class and the product group.

These selection criteria were selected because experience has shown that the grouping of single parts into design families cannot be made according to the shape alone, but — in addition to the actual parts function — the product function and requirements have also to be considered, because they have a decisive influence on the design and machining aspects of the part.

This mode of selection has proved to be suitable and it also simplifies the subsequent standardization process because the product influences are excluded.

Next a classification according to shape and material is offered, whereby the so-called similarity field, which is formed from the workpiece code, has proved to be useful. As a result of this rough classification, a so-called information journal of the determined single part drawings is compiled from the increasing workpiece code number. This also shows the part designation as well as the article and drawing identification.

Together with a visual perusal of drawings, a pre-grouping into design families is made from the shape and part function standpoint. This preparatory work stage is a part of the selection phase as it is not yet based on any analysis and it should serve as a limitation of the investigation field because, in the process of standardization, certain adjustments will be necessary.

The frequency distribution of the design families serves as a basis for the prognosis in respect of the expected employment frequency of the similar parts and is decisive for the selection of the rationalization method in the sense of effort and return. The frequency distributions shown in Fig. 6.4 — even if influential — only represent the classification criteria, because further influential factors such as part complexity, drawing effort, standardization effort and the effects of the rationalization methods in manufacture have to be considered as well.

6.2.2 Analysis

The analysis (Fig. 6.5), for which the three defined similarity stages constitute the basis, has the principle objective of determining all the parts of the respective design family. This is effected by the fact that in stage III, those drawings, which have a high degree of similarity in all three characteristics, are analysed and grouped together. On the basis of this grouping, a frequency diagramm is drawn up according to the number of drawings and a minimum number of pieces determined. The respective similarity group represents the starting point for the stage II analysis. Stage II provides for analysis of the single part drawings according to the frequency of similar basic shapes and elements, and the non-considered parts are then used as a starting basis for stage I. The remaining parts are analysed according to the frequency of their elements. In conjunction with the frequency limit for the individual standardization stages, this step-by-step analysis enables the crucial points of rationalization to be simply evaluated and a systematic starting point to be created for the synthesis.

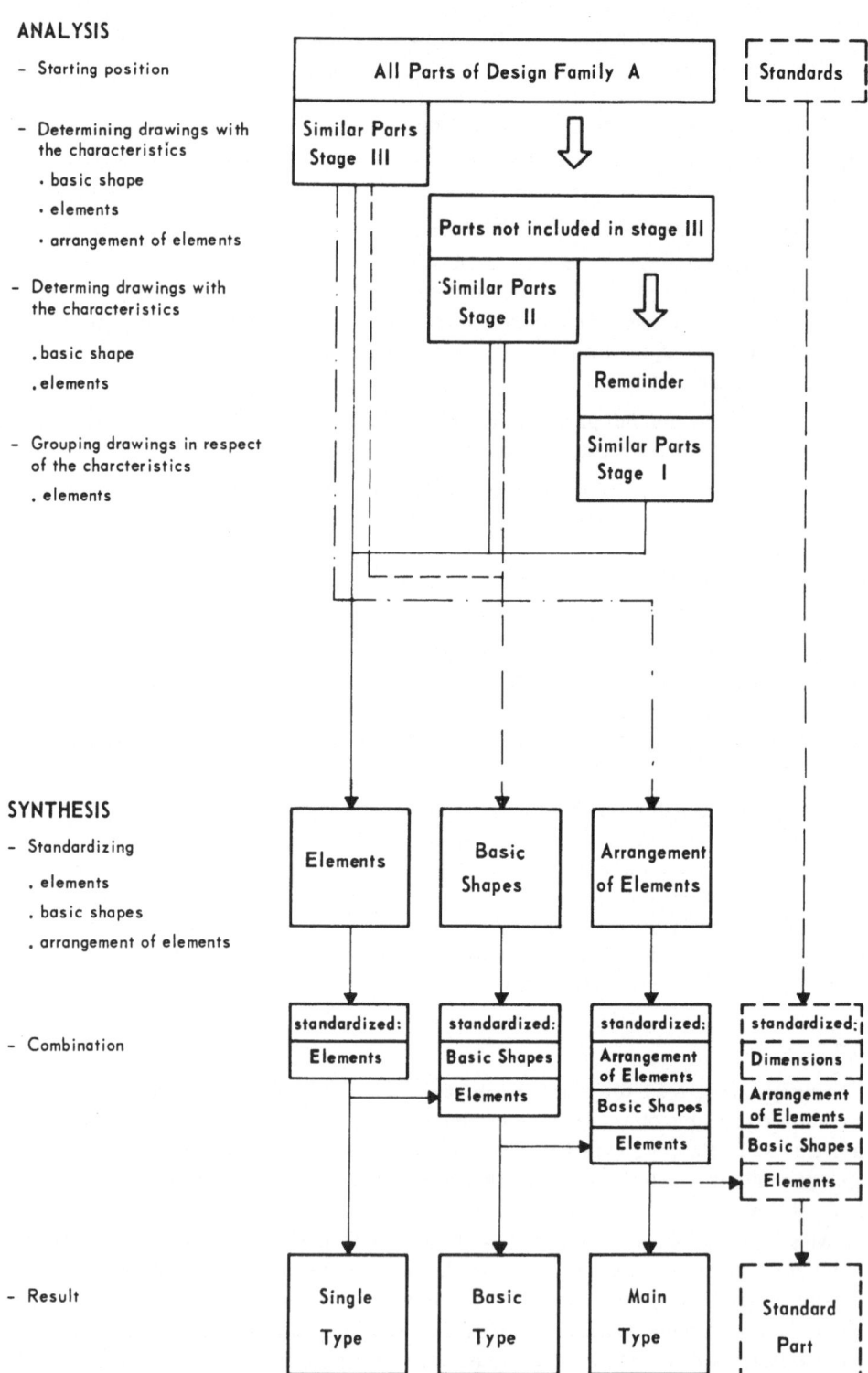

Fig. 6.5. Relationship between Analysis and Synthesis within the Procedure for Developing Similarity Types

6.2.3 Synthesis

The synthesis is based on the previous analysis (Fig. 6.5). In the first step, the elements of all three similarity stages are duly considered. During the standardization of elements, care should be taken that the main emphasis falls more on the elements of stage III (main type) than those of stage I and II, because they assign to a parts spectrum with a higher degree of standardization. This is also effected in an analogous manner with the standardization of the basic shape which is based on the analysis of stages II and III.

The elements, basic shapes and arrangements of elements standardized in this way, constitute the basis for developing single, basic and main types.

6.2.4 Working Aids

The similarity data, which comprises the main, basic and single types of various design families, provides the working aids for its users. An important requirement concerning the configuration of the catalogue is the provision of a well-ordered index system which will enable the required information to be found within a short time.

The catalogue is built up in such a manner, that, within a design family, first the highest similarity stage with main types can be found. If those are not able to meet the specific requirements, the next lower stage with its basic types and greater freedom of design becomes avaiable. Should this stage still not be able to provide the solution, then the final stage of single types — standardized elements — can be used.

6.3 Model Example: Similarity Types of the Design Family

Selection

A design family selected from the pump assortment — explained in Fig. 5.6 — in accordance with the following selection criteria is used as a model example (Fig. 6.6):

Product class:
— hydraulic machines and plants;

Product group:
— high-pressure centrifugal pumps,
— low-pressure centrifugal pumps,
— bore hole pumps;

Similarity fields:
— shape field 00 (cutting process, rotary parts without deviation, disc-shaped and flanges without threads),
— material fields 1, 2 and 3 (1: steel without heat treatment, 2: cast and malleable iron, 3: Al and Cu-alloys).

With the investigated product spectrums, all the workpieces were already available in coded form. Using an evaluation programme as an aid, the required parts were sorted from the data bank and listed in an information journal according to their increasing workpiece code number. On examining the respective drawings, it became evident that the design family 'A' was a promising one. It contained 205 parts, which had the same main shape and the same basic function, namely to prevent wear on casing and impeller.

SELECTION

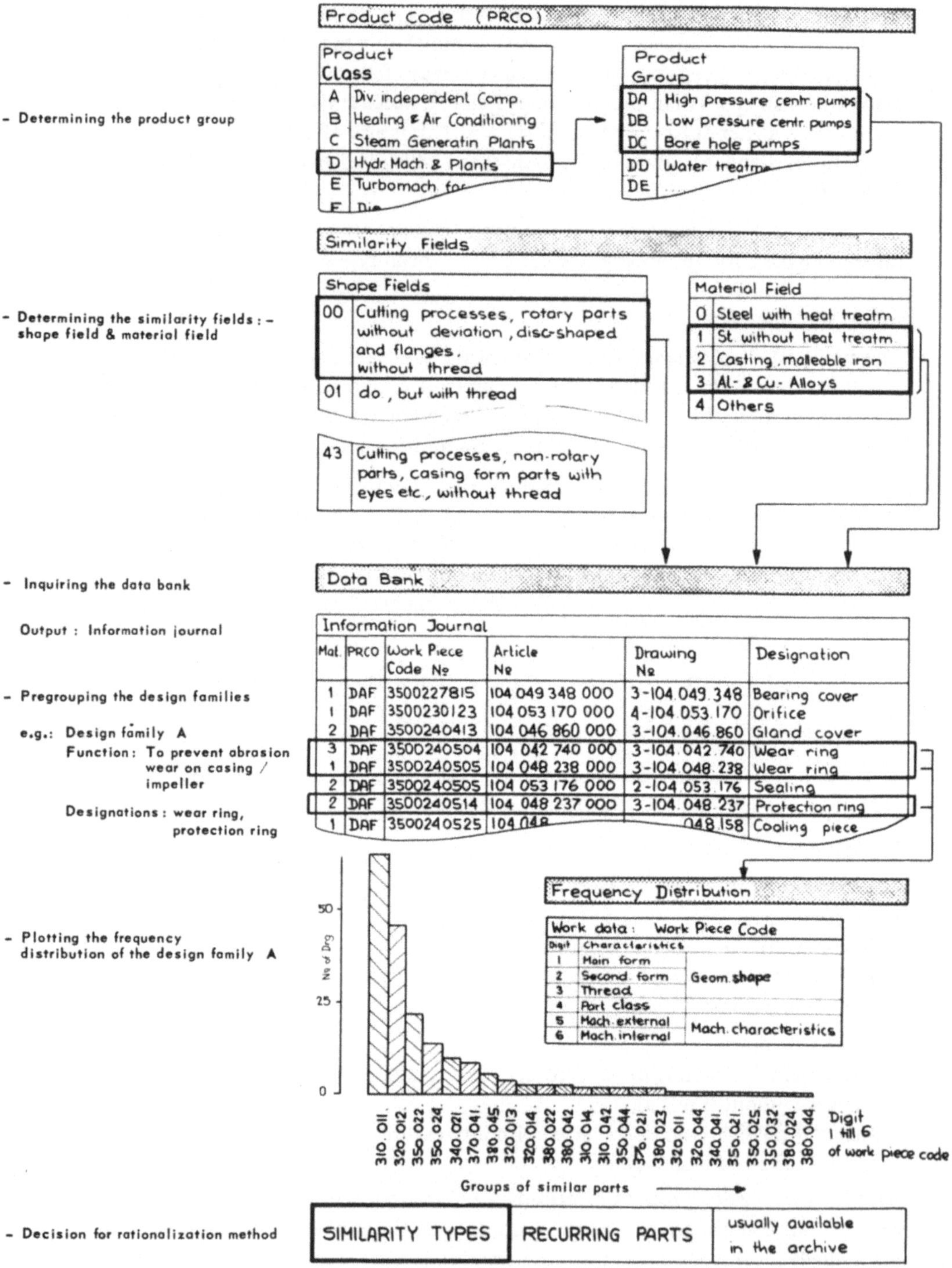

- Determining the product group

Product Code (PRCO)

Product Class			Product Group	
A	Div. independent Comp.		DA	High pressure centr. pumps
B	Heating & Air Conditioning		DB	Low pressure centr. pumps
C	Steam Generatin Plants		DC	Bore hole pumps
D	Hydr. Mach. & Plants		DD	Water treatm
E	Turbomach for		DE	
F	Dia			

- Determining the similarity fields : –
  shape field & material field

Similarity Fields

Shape Fields		Material Field	
00	Cutting processes, rotary parts without deviation, disc-shaped and flanges, without thread	0	Steel with heat treatm
		1	St. without heat treatm
01	do, but with thread	2	Casting, malleable iron
		3	Al- & Cu- Alloys
43	Cutting processes, non-rotary parts, casing form parts with eyes etc., without thread	4	Others

- Inquiring the data bank

Data Bank

Output : Information journal

- Pregrouping the design families

  e.g.: Design family A
  Function: To prevent abrasion wear on casing / impeller

  Designations : wear ring, protection ring

Information Journal

Mat.	PRCO	Work Piece Code №	Article №	Drawing №	Designation
1	DAF	3500227815	104 049 348 000	3-104.049.348	Bearing cover
1	DAF	3500230123	104 053 170 000	4-104.053.170	Orifice
2	DAF	3500240413	104 046 860 000	3-104.046.860	Gland cover
3	DAF	3500240504	104 042 740 000	3-104.042.740	Wear ring
1	DAF	3500240505	104 048 238 000	3-104.048.238	Wear ring
2	DAF	3500240505	104 053 176 000	2-104.053.176	Sealing
2	DAF	3500240514	104 048 237 000	3-104.048.237	Protection ring
1	DAF	3500240525	104 048	048.158	Cooling piece

- Plotting the frequency distribution of the design family A

Frequency Distribution

Work data : Work Piece Code

Digit	Characteristics	
1	Main form	Geom. shape
2	Second. form	
3	Thread	
4	Part class	
5	Mach. external	Mach. characteristics
6	Mach. internal	

N° of Dng

Groups of similar parts ⟶

Digit 1 till 6 of work piece code

- Decision for rationalization method

SIMILARITY TYPES	RECURRING PARTS	usually available in the archive

Fig. 6.6. Model Example: Selection

86

This example shows quite clearly that sorting according to the designation would not lead to a result on its own, because the various terms such as casing wear ring, impeller wear ring, rotating seal ring and sealing ring are used for the same application. When we consider that the multiplicity of an international group is greatly influenced by the language factor, then the need for a language independent classification system is even more emphasized.

On the basis of the characteristic 'main shape' within the design family 'A' under investigation, a frequency diagramm was prepared for the selection of the rationalization method. Within the framework of this design family, a limit of four workpiece drawings approximately represents the break-even point between the standardization effort and the future savings. This covers the simplified drawing work and the resultant more rational process planning as well as the indirect effects on manufacture due to the narrowing down of the parts variety and the creation of conditions for the application of equipment for computer aided design. In this particular case, the choice led to similarity types.

Analysis

The starting point for the analysis was determined by the three characteristics — basic shape, elements and arrangement of elements. These were applied to the investigated design family in a figurative way and are shown in Fig. 6.7 in an extracted form.

The first step was to examine the drawings for the similar combinations of the three characteristics that had occured the most frequently. In doing so, the minimum number of four drawings was again selected so as to assign a combination as main type to stage III. From the remaining parts, the similar basic shape shown in the middle of Fig 6.7 is determined as the assignment criteria for stage II. The remaining drawings again constituted the starting point for stage I. Within the framework of this design family, further elements (e.g. drill holes) were analysed and laid down as the assignment criteria.

This investigation showed that about 8/12 of all the parts could be assigned to stage III, 3/12 to stage II and 1/12 to stage I. The reason for this is that a relatively large number of simple parts belongs to this design family. With complex parts, the distribution would be displaced in the direction of stage II and I.

Synthesis

The following synthesis was carried out, based on the result of the analysis (Fig. 6.8). In a first stage, the elements of all three stages were grouped together and listed in a tabular summary in accordance with application frequency. The elements intended for standardization were analysed according to their specific characteristics and grouped in similarity classes, e.g. in the case of the elements 'hole for countersunk screw 90°' according to the

— Application area (ring diameter),
— Number per circumference,
— Hole diameter.

The resultant frequency in the illustrated matrix (Fig. 6.8), thus constitutes the basis for the standardization proposals for the single types.

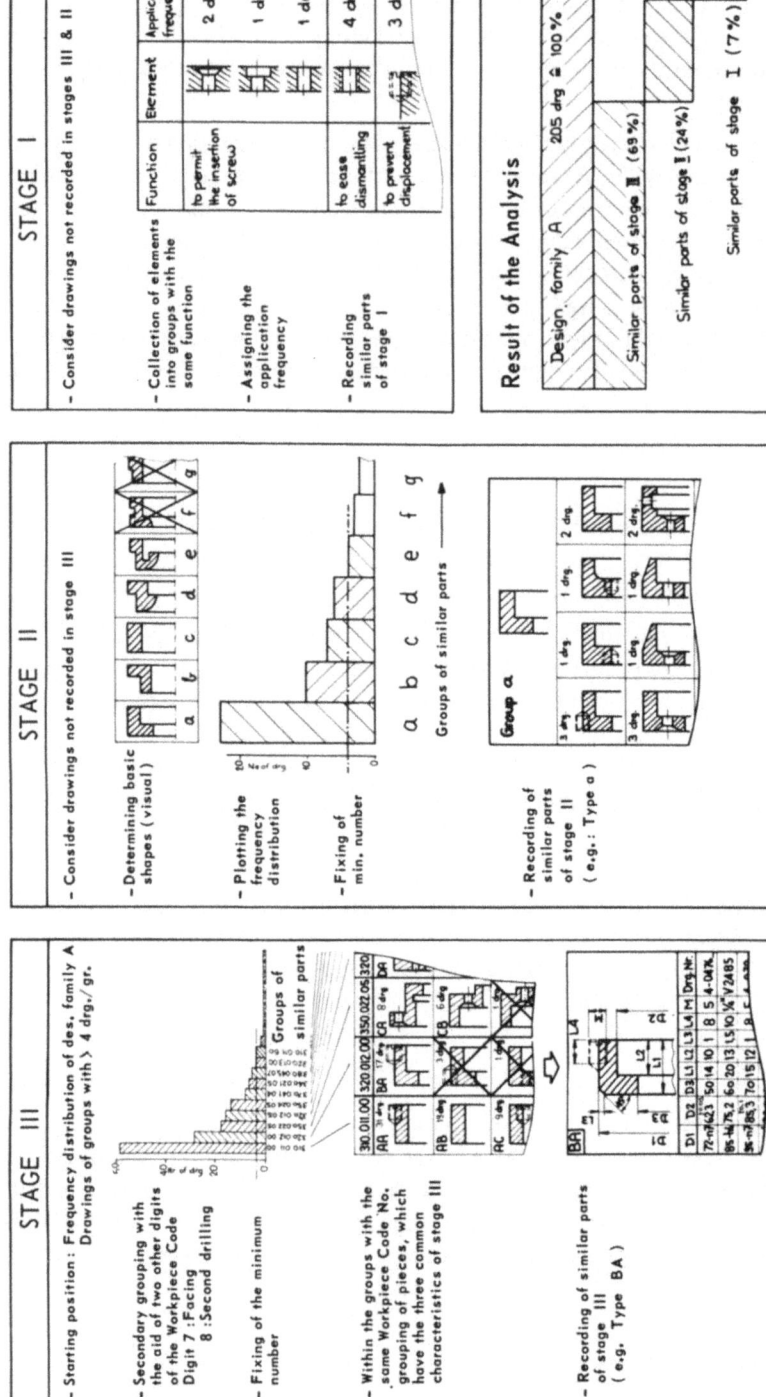

Fig. 6.7. Model Example: Analysis

SYNTHESIS

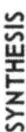

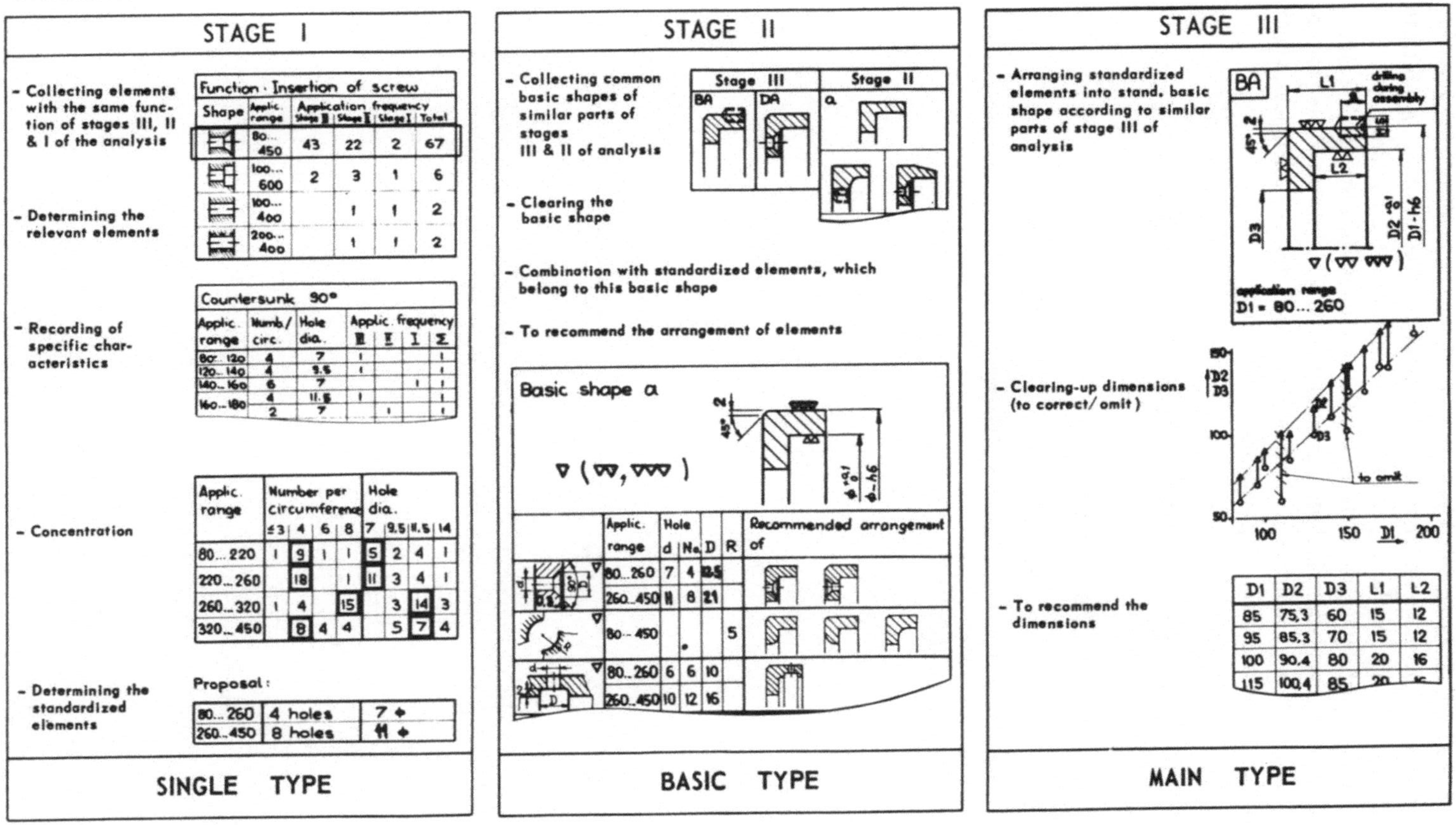

Fig. 6.8. Model Example: Synthesis

This process was repeated in an analogous manner for the basic and main types. A further step was added, however, whereby — apart from the standardized characteristics — recommendations were made under which a degree of freedom was left open intentionally for the parts design. As the degree of freedom decreases, the similarity type becomes a standard part, by which all the characteristics identifying the part are laid down. These recommendations include, for example, the arrangement of elements of technical manufacturing requirements such as dimensional areas or tolerances, which are orientated to the manufacturing systems.

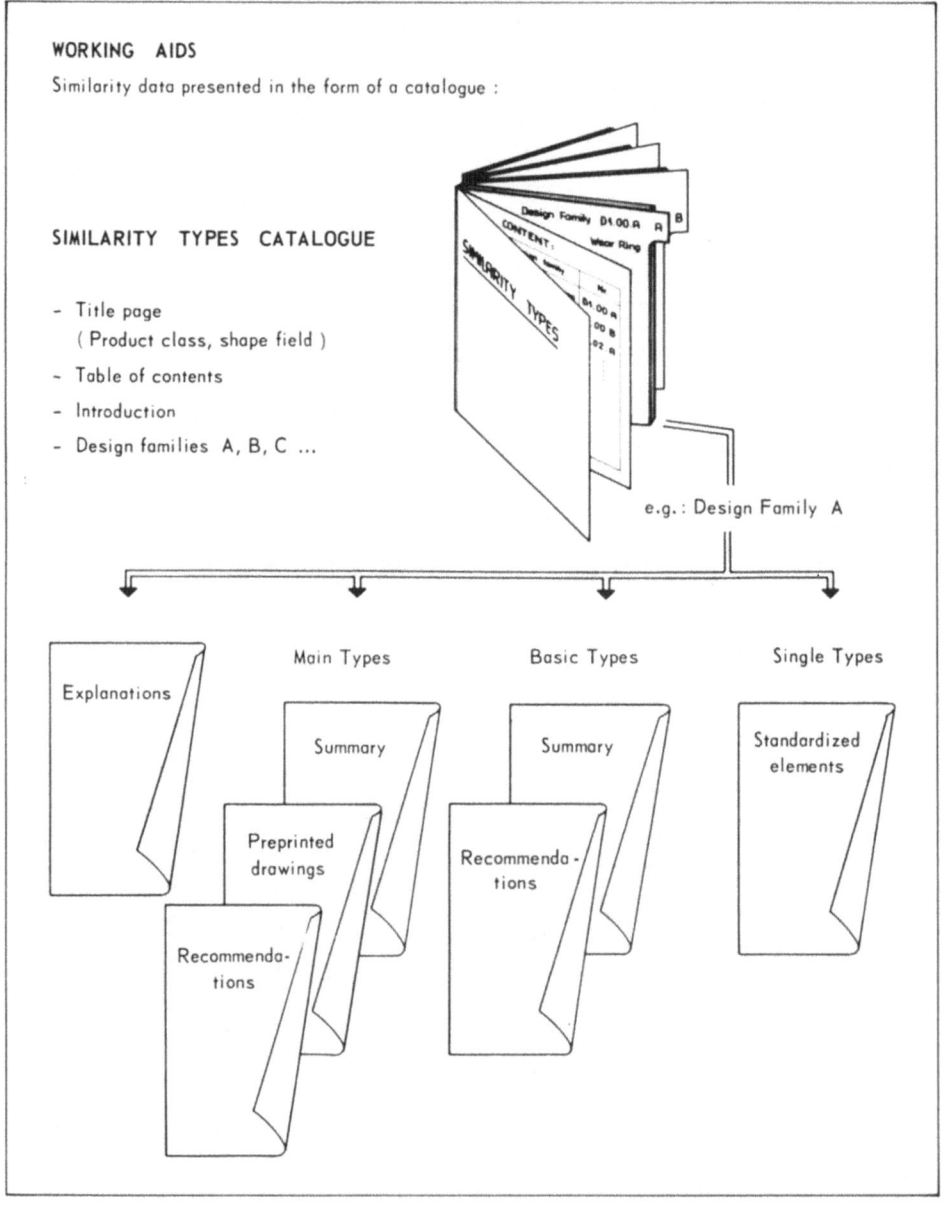

Fig. 6.9. Model Example: Working Aids

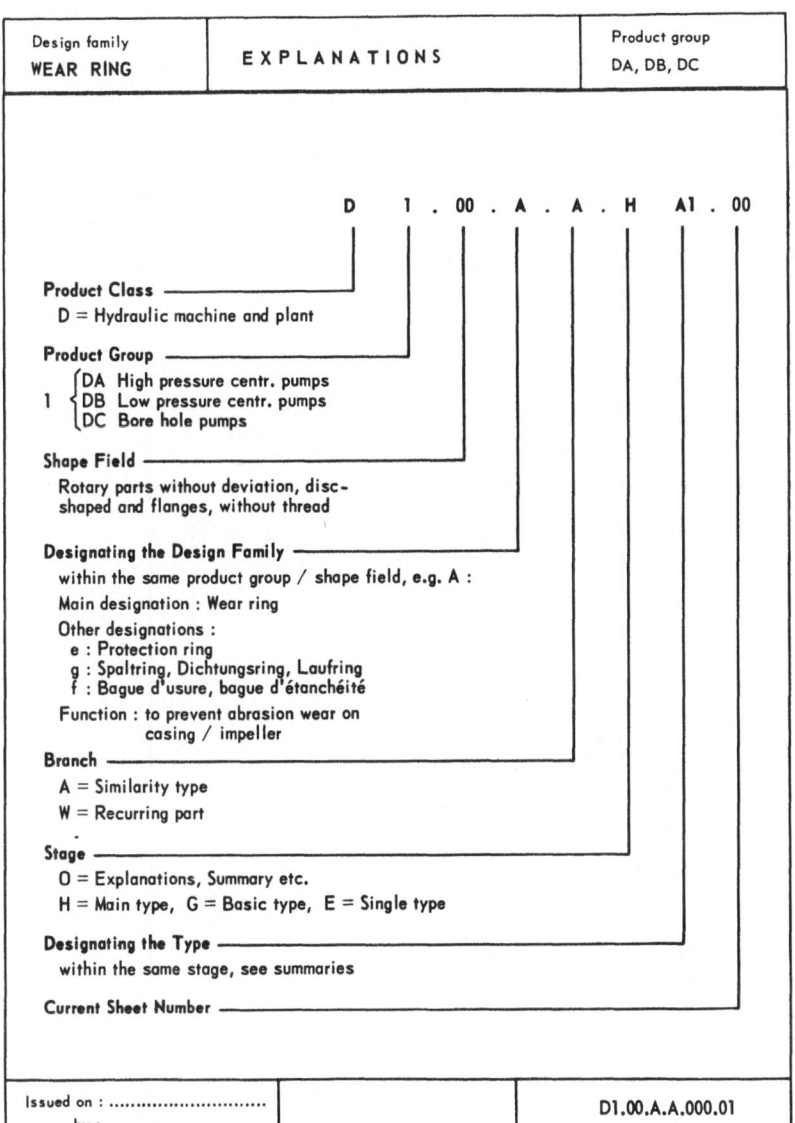

Design family	EXPLANATIONS	Product group
WEAR RING		DA, DB, DC

D 1 . 00 . A . A . H A1 . 00

Product Class ─────────
D = Hydraulic machine and plant

Product Group ─────────
  ⎧ DA High pressure centr. pumps
1 ⎨ DB Low pressure centr. pumps
  ⎩ DC Bore hole pumps

Shape Field ─────────
Rotary parts without deviation, disc-
shaped and flanges, without thread

Designating the Design Family ─────────
within the same product group / shape field, e.g. A :
Main designation : Wear ring
Other designations :
  e : Protection ring
  g : Spaltring, Dichtungsring, Laufring
  f : Bague d'usure, bague d'étanchéité
Function : to prevent abrasion wear on
         casing / impeller

Branch ─────────
A = Similarity type
W = Recurring part

Stage ─────────
O = Explanations, Summary etc.
H = Main type, G = Basic type, E = Single type

Designating the Type ─────────
within the same stage, see summaries

Current Sheet Number ─────────

Issued on :		D1.00.A.A.000.01
by :		

Fig. 6.10. Model Example: Explanations of Design Family A

An absolute percentual result cannot be provided in the case of the standardized parts because the basic shapes and elements of the single parts were already considered during the standardization in all three stages. Nevertheless, the following numbers of similarity types were determined in this design family:

— Main types: 12,
— Basic types: 5,
— Single types (various elements): 8.

Working Aids

On the basis of these results, similarity data were prepared and are shown in Fig. 6.9. The first sheet (Fig. 6.10) explains the index system. The most frequent designation term is used as the term for identification. Further explanations are given in respect of the shape description, the function and the remaining designations. An example is used to explain the structure of the identification number for the parts included in this design family. The next sheet (Fig. 6.11) is used to summarize the main types. The individual main types are shown in the form of pre-printed drawings (Fig. 6.12) and supplemented by a sheet of recommendations (Fig. 6.13).

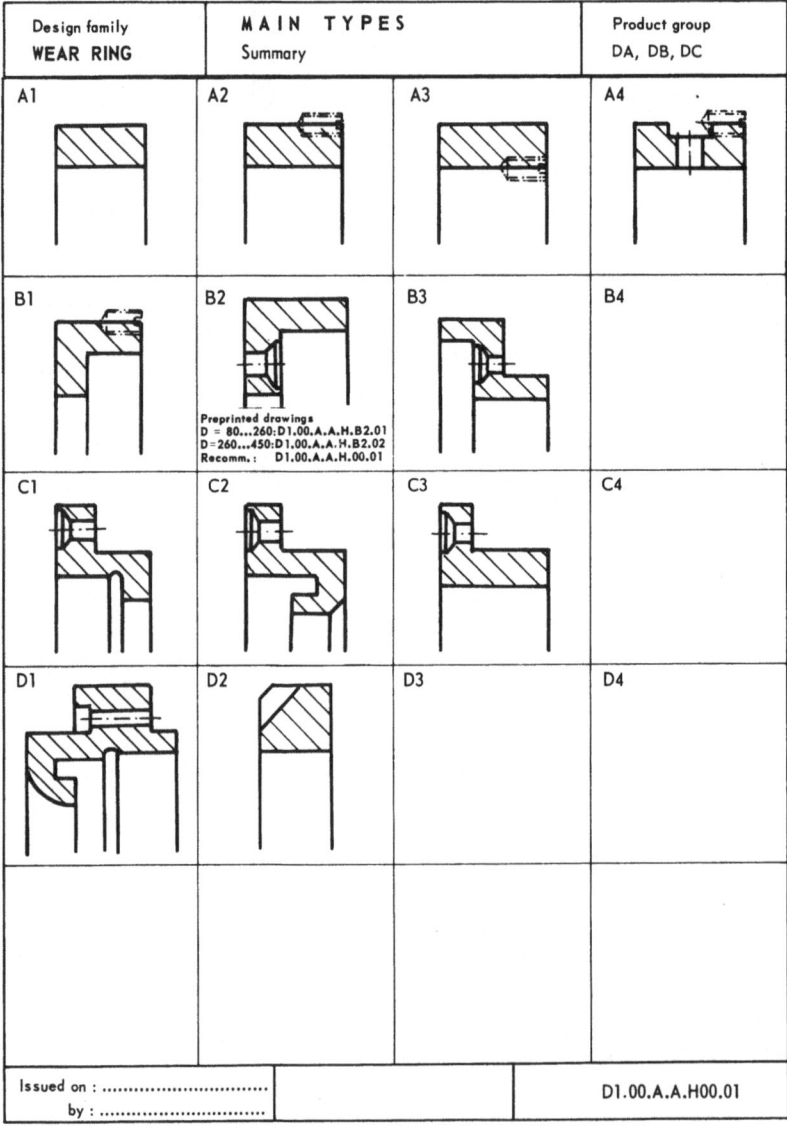

Fig. 6.11. Model Example: Summary of Main Types

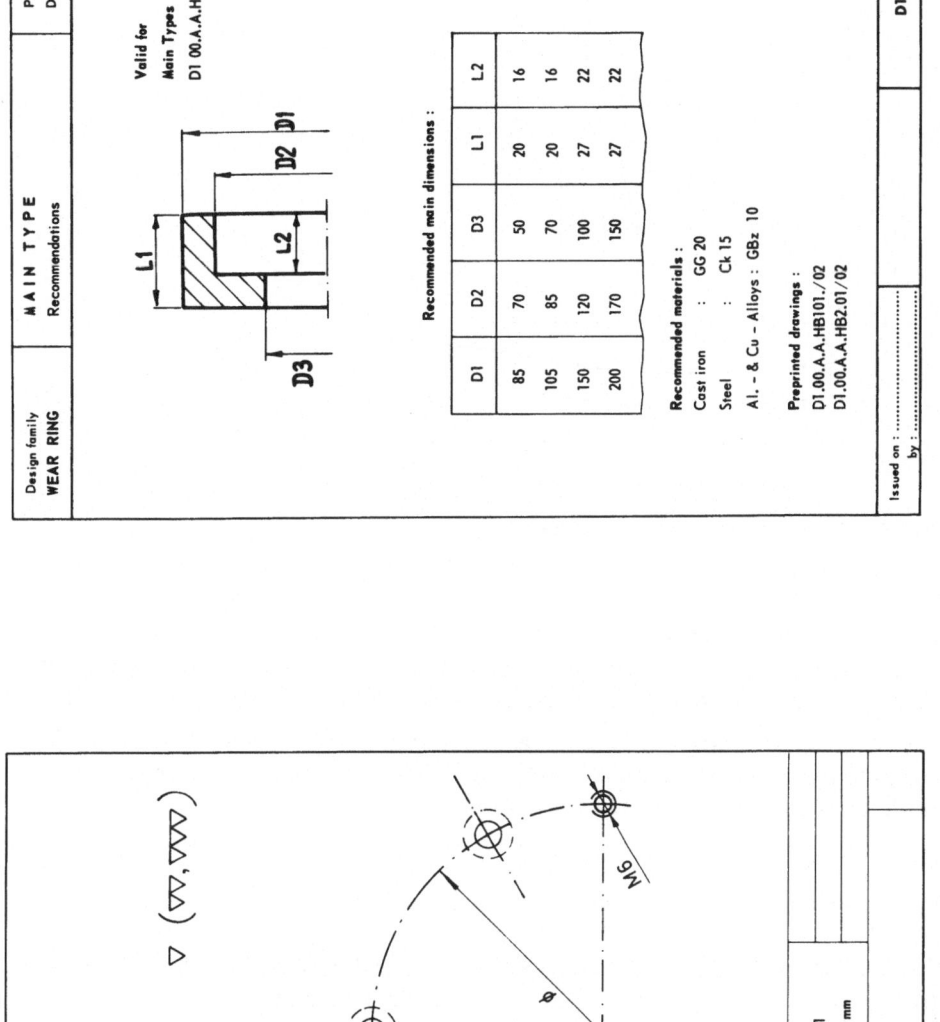

**Valid for
Main Types**
D1 00.A.A.HB1/2

Recommended main dimensions :

D1	D2	D3	L1	L2
85	70	50	20	16
105	85	70	20	16
150	120	100	27	22
200	170	150	27	22

Recommended materials :
Cast iron : GG 20
Steel : Ck 15
Al. – & Cu – Alloys : GBz 10

Preprinted drawings :
D1.00.A.A.HB101./02
D1.00.A.A.HB2.01/02

Issued on : ....
by : ....

D1.00.A.A.H00.01

Fig. 6.13. Model Example: Recommendations for a Main Type

Preprinted drawing No. D1.00.A.A.HB2.01, valid for outer diameter range 80... 260 mm

WEAR RING   Main type D1.00.A.A.HB2.01
Valid for outer dia. 80... 260 mm

Fig. 6.12. Model Example: Preprinted Drawing of Main Type

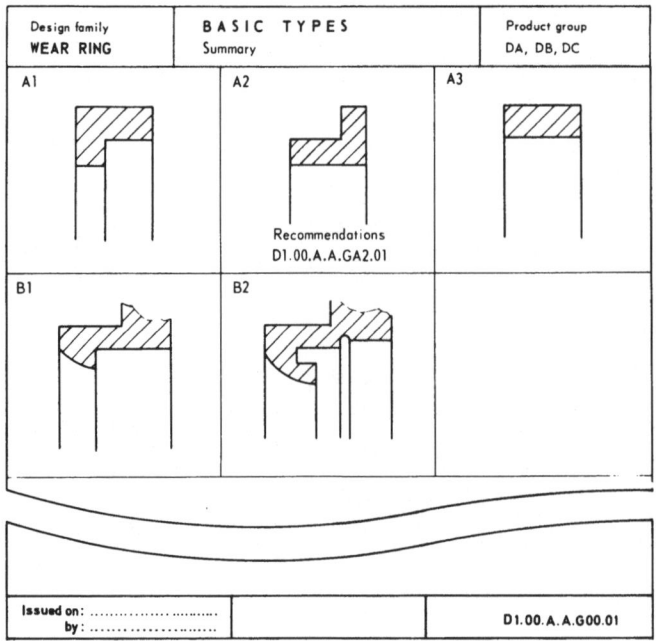

Design family WEAR RING	BASIC TYPES Summary		Product group DA, DB, DC
A1	A2 Recommendations D1.00.A.A.GA2.01	A3	
B1	B2		

| Issued on: .... by: .... | | | D1.00.A.A.G00.01 |

Fig. 6.14. Model Example: Summary of Basic Types

An analogous procedure is also employed for the basic types (Fig. 6.14). This supplement sheet (Fig. 6.15) includes the recommended elements and their arrangement. Fig. 6.16 shows a section of the standardized elements for the single types.

The work sheets are arranged in accordance with the sequence stage III, II and I. The incorporation of the working aid should be effected in accordance with the existing standards organization of the respective firm.

6.4 Aspects of Computer Aided Parts Design

The described method for the rationalization of parts design on the basis of GT constitutes a good foundation for further rationalization efforts in the direction of computer aided design. The similarity data can be used directly as a basis. Contrary to a general programme, which enables a great number of possible cases to be covered and necessitates very considerable development effort, the proposed solution permits a self-contained programme to be built up, which is based on a prognostic application frequency. In this direction, the work performed by Steinmetz [37] shows a possible approach for computer aided design.

Design family WEAR RING	BASIC TYPE D1.00.A.A.GA2. Recommendations	Product group DA, DB, DC

Elements	Application range (do, Di)	numb./circ.	Dimensions d	D	B	Recommended arrangement
	80...260	4	7	12,5		
	260...450	8	11	21		
	80...260	2	M6			
	260...450	2	M10			
	190...600					

Recommended materials :

Cast iron : GG 20
Steel : Ck 15
AL.-& Cu-Alloys : GBz 10

Execution of Drawings :

Manual to draw : To draw
With computer aid : Equipment .... / Progr. No. ....

Issued on : .... by :		D1.00.A.A.GA2.01

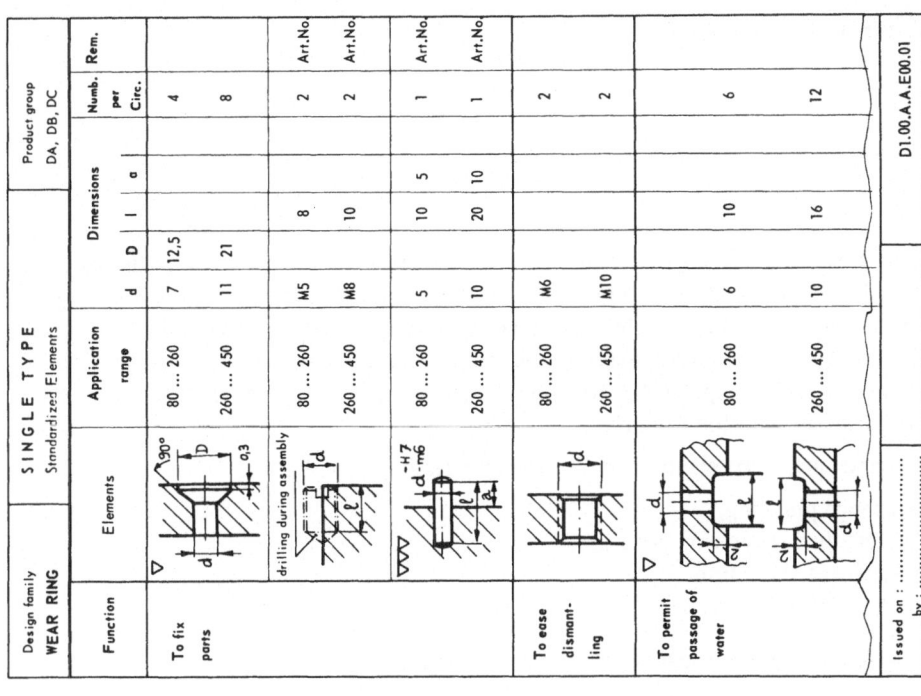

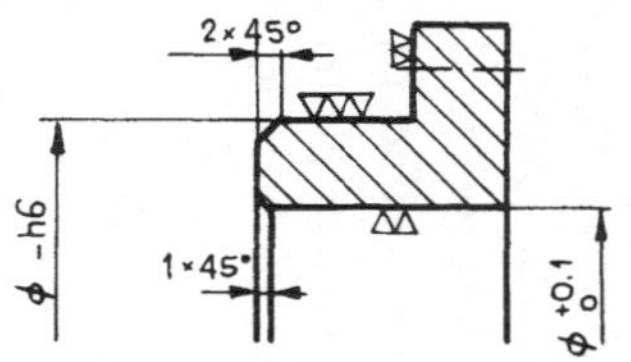

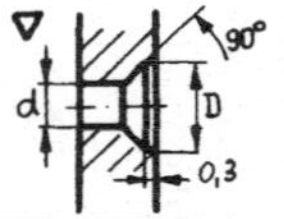

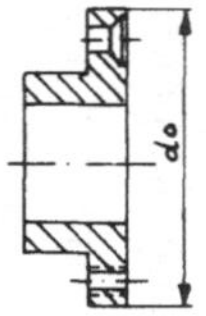

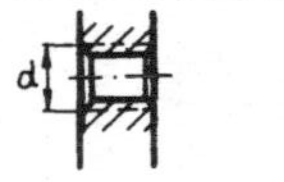

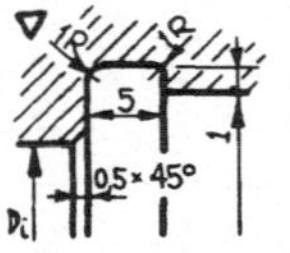

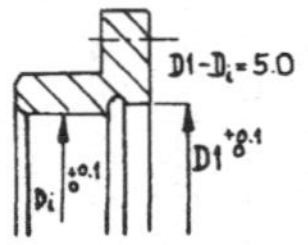

Fig. 6.15. Model Example: Recommendations for a Basic Type

Design family WEAR RING	SINGLE TYPE Standardized Elements	Product group DA, DB, DC

Function	Elements	Application range	Dimensions d	D	l	a	Numb. per Circ.	Rem.
To fix parts		80 ... 260	7	12,5			4	
		260 ... 450	11	21			8	
	drilling during assembly	80 ... 260	M5		8		2	Art.No.
		260 ... 450	M8		10		2	Art.No.
		80 ... 260	5		10	5	1	Art.No.
		260 ... 450	10		20	10	1	Art.No.
To ease dismantling		80 ... 260	M6				2	
		260 ... 450	M10				2	
To permit passage of water		80 ... 260	6		10		6	
		260 ... 450	10		16		12	

Issued on : .... by :		D1.00.A.A.E00.01

Fig. 6.16. Model Example: Standardized Elements for Single Types

7 Process Planning

In mechanical engineering the term 'production process' infers all the activities associated with the manufacture of a product, i.e. from the design stage, process planning and manufacture to the assembly. In process planning, which represents the most important link between development and manufacture, the drawings and parts lists prepared by the design department are converted into machining-orientated instructions in the form of operational charts.

The operational chart elaboration includes the following activities:

— Sequence planning,
— Manufacturing system allocation,
— Equipment allocation,
— Work measurement.

The work measurement will be dealt with in Chapter 8.

More refined working methods, which make possible a more exact planning, have been applied in the production of large batches for some time. On the other hand, conventional methods and work techniques are still applied in a large number of firms for one off and small-batch production. These methods and techniques, however, only make planning feasible in an inexact manner and with a small degree of assurance. When the relationship of the effort between process planning and manufacturing — in relation to the individual workpiece — is very close, then this crucial point of rationalization becomes a matter of particular importance. This fact is accentuated by the following investigation which has been carried out in twelve mechanical engineering firms using a machined rotary part (Fig. 7.1) on the basis of various batch sizes [3].

When the direct time of process planning is related to a workpiece of one-off or small batch production, we find that it is higher than that in respect of a workpiece of medium or large batch production. When considering the situation we should also make due allowance for the frequency of the planning orders over a certain period of time because this increases the difference between small, medium and large batch production. Apart from the planning variety, we should not forget that the shortage of personnel and personnel changes are increasing in process planning and consequently the potential of average experience is declining. This also has a detrimental effect on the planning quality and the throughput time. When we consider that process planning provides the basis for the offer and cost calculation work, the necessity of decision data is particulary emphasized.

We are already familiar with fundamentals, methods and aids of process planning from literature and practice [1, 2, 4, 7, 38, 39, 40, 41, 42]. Generally speaking, we can differentiate between new planning and similarity planning. Concerning new planning, each case is built up synthetically with the aid of the decision data, as is practised analogously with predetermined motion time systems. The new planning is characterized by its wide application range and flexibility. Nevertheless, the system structure calls

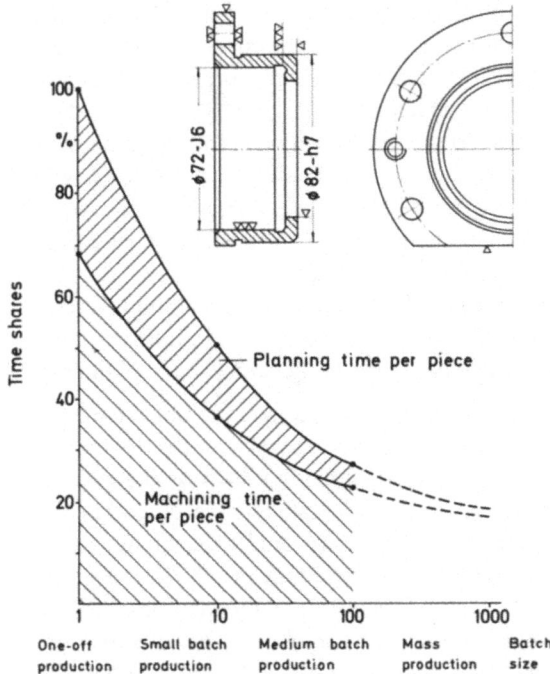

Fig. 7.1. Planning and Machining Times in Relation to Different Batch Sizes

for a large deal of information and rational processing of the data. The similarity planning is based on the comparison principle, by which workpieces with similar machining characteristics are grouped together in a so-called 'machining family' and from which similarity data is built up. This is then used as a comparative basis for the planning of similar machining tasks. The system structure is simple and can be extended stepwise. GT and the classification system thereby constitutes a rational planning basis.

As with the parts design, three-staged similarity data are again selected to ensure a flexible and economic structure. As a result of the frequency and the degree of similarity of the similar parts, these three stages differentiate in the degree of detailing of the similarity data. The terms 'main', 'basic' and 'single' types are again used to define the three similarity stages.

In the field of one-off and small batch production, various classification systems are already known with the aim of structuring the parts spectrum in similar workpiece groups.

As already mentioned in detail in Chapter 3 and 4, two general approaches are available for building up the classification system. These differ in the creation of a specific or general planning basis. In this concept, the general classification system described in Chapter 4 is used as a basis for describing the workpiece shape, the operations and the equipment. The investigations showed that with the aid of specific classification systems, it is only possible to obtain a detailed planning basis, but the subsequent standardization effort is in no way reduced.

The classification systems only have a preparatory function and so it is quite sufficient to operate with fewer but more relevant similarity criteria in this particular phase.

The functional relationship of the production process shows that a direct relation exists between parts design and process planning. This allows the conclusion that two problem circles – parts design and process planning – may be incorporated in the same rationalization process. The analysis of the problem indicates, however, that the build-up should be effected separately according to the design and manufacture application areas. The term 'parts family', which is referred to in the literature, will be retained as an overall term. For the structure according to the two application areas, the term 'design family' described in Chapter 6 and 'machining family' which is applied as a basis for process planning, are used. This enables the two partial areas to be built up in a more simple and application-orientated manner. The data, however, should be matched to each other on their marginal lines with the purpose of reciprocal influence. This appears to be expedient in view of a later integrated parts design and process planning.

In opposition to technical investment and layout planning, where the objective is to find an optimum manufacturing system for a specific parts spectrum, the decision process in the case of process planning runs in an opposite direction, because the manufacturing systems are fixed [39]. The task is therefore to allocate technically and in an optimum manner a certain parts spectrum with the same or similar machining operations to the manufacturing systems. The reference to the connection indicates quite clearly that the objective of the similarity methods is not to be directed towards the individual case but to the consideration of an individual machining family with similar complete machining. The aspect of overall throughput time has therefore to be included in the workpiece allocation within the framework of the process planning and especially considered in the GT planning principles.

Using this situation analysis as a starting point, a solution is to be found which places the entire problem complex in a comprehensive framework. The flexible application of a process planning system calls for stepwise rationalization in the sense of effort and return.

The three-stage similarity data are based on representative similar parts which form a machining family. By a similar part, we understand a workpiece which, based on its criteria, can be allocated to a similarity stage of a machining family. The machining families are limited by the allocated design families and the machining groups of the individual GT-manufacturing systems. The following crucial points can be regarded as the aspired objectives:

– The rationalization reserves of process planning should be fully utilized with the aid of standardized decision data on the basis of GT.
– The machining families should be presented by the similarity data in such a way that they allow the similar parts to be allocated quickly and economically with the aid of the shape and machining characteristics.
– The machining families should be able to be manufactured with the optimum manufacturing systems and the corresponding limited assortment of machining aids with the principal objective to minimize the total throughput time for the complete manufacture.
– The elaborated similarity data should constitute a basis for the automatic preparation of the operational charts.
– The NC-machines, however, are a special case. The similarity data here includes the NC-programming families for simplifying the programming work and is principally aimed at improving the economic application of NC-machines.

7.1 Similarity Planning

In this section, the interrelationship between the planning of a manufacturing task and the use of working aids is shown in Fig. 7.2 which is a schematic representation of the planning procedure. The allocation of a workpiece to the corresponding machining family is effected on the basis of rough planning for the complete machining. In an analogous manner as in the parts design, the product groups and parts types with similar product requirements were selected as superimposed criteria for allocation. The limitation of the machining families is effected with the aid of similarity fields.

In a further step, the working operations and the operation sequence are established. This step also provides for the allocation to a suitable similarity type which is then used as a comparative basis for the fine planning of the individual working operations. The third step is used to establish the detailed machining conditions, the optimum manufacturing system and the individual machines.

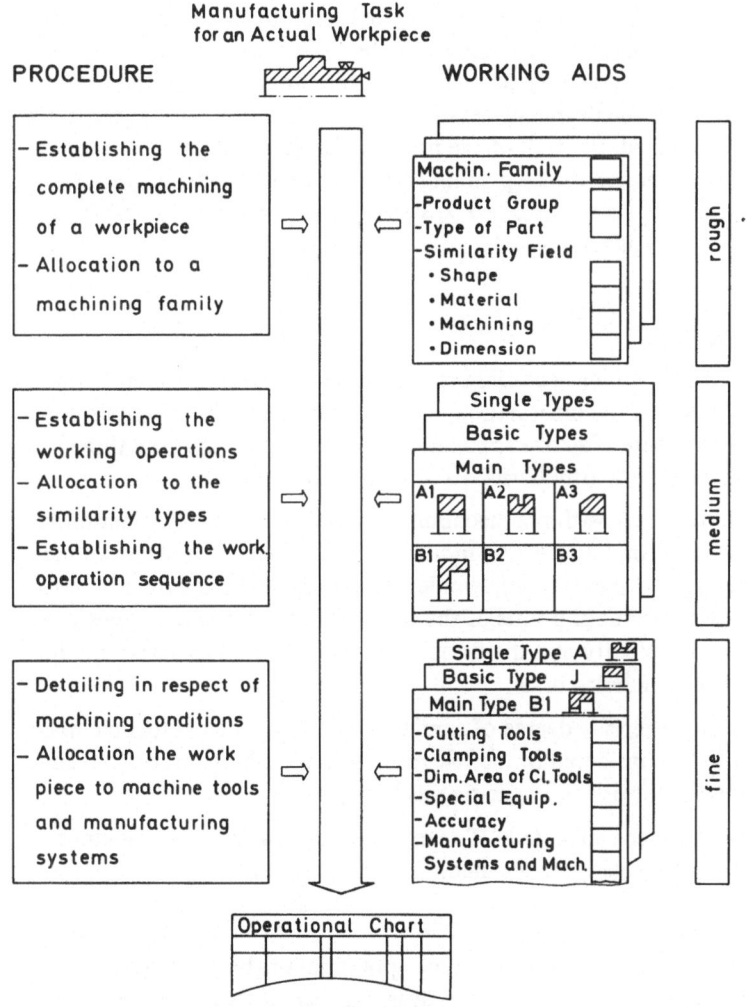

Fig. 7.2. Procedure Plan for Similarity Planning

The degree of detailing of the similarity data is derived from the frequency of previous representative similar parts and their degree of similarity in the sense of a prognosis concerning the frequency of application. On the other hand, the degree of detailing of the individual manufacturing tasks is dependent on further factors such as the degree of automation of the machines as well as the average annual batch size and the personnel structure (training and experience potential of the personnel). This procedure has been selected as a result of economic considerations. Detailed decision data should only be prepared in advance, when the advantage of application is assured. This particular matter is discussed in a detailed manner in section 7.2. Using the similarity data as a basis, the detailed working operation description can now be elaborated and the optimum machine selected but this must be effected in accordance with the superimposed GT-manufacturing system. The alternative machines in the similarity data are thus listed according to the sequence GT-line, cell and centre. The determination is made in an iterative process.

The similarity data for the single types only provide general indications concerning the necessary manufacturing aids for the working operation. The basic types provide additional information as to the special equipment, tools etc., the main types, however, describe the working operation in larger detail using the 'partial operation' as a basis. This procedure is also used in an analogous manner for the programme build-up for NC-machines. Depending on the degree of detailing for the similarity data or respectively the workpiece to be planned, the task of the process planner consists in comparing data. From this results a reduction of the planning effort and enables the degree of similarity for the complete parts spectrum to be steadily increased. Thus a basis for the next rationalization stage is established.

7.1.1 Similarity Types

The focal point of process planning according to the similarity method is constituted by the working aids, by which the information is included in the form of various similarity types. In order to ensure effective application, it is important that the three-staged similarity types be expediently harmonized with each other.

The selection of similarity types within a machining family is — as already mentioned — dependent on the degree of similarity of the individual similar parts. The relevant machining criteria of the respective machining families are used to determine the degree of similarity. Fig. 7.3 shows the crucial points of the three different similarity types as a function of the degree of similarity and the frequency, which is used to determine the degree of detailing for the similarity data.

Apart from the two main criteria — degree of similarity and frequency — further marginal conditions have to be considered in the selection of the similarity stages and similarity types within the framework of a specific machining family, e.g.

— Degree of automation for the allocated GT-manufacturing systems and machines,
— Complexity of the workpieces to be planned,
— Effects on quality,
— Annual number of pieces manufactured,
— Personell structure (training level and experience potential of the personel),
— Effects on the long-term rationalization objectives, e.g. computer-assisted process planning.

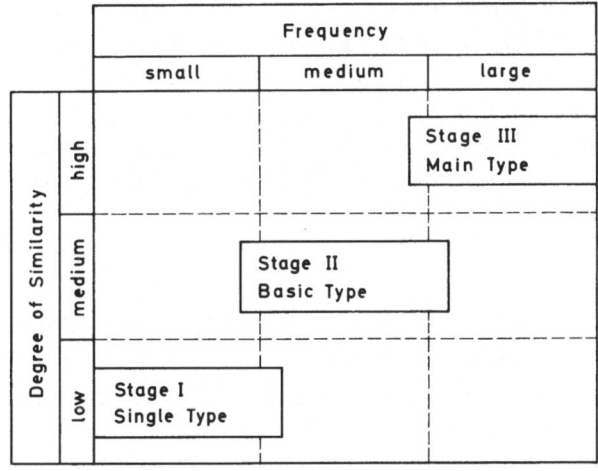

Fig. 7.3. Major Determination Factors for Similarity Types

In principle, it can be concluded that with an increasing degree of detailing the technical-economic success will be improved in the direction of optimum process planning, but the effort in respect of building up the similarity data will increase at the same time. The limit of frequency has to be established for every similarity stage on the basis of this criteria. In doing so, care should be taken that not only the direct effects, such as the direct cost reduction per operational chart, but the indirect effects in the rationalization of manufacture are also considered. A high degree of detailing for the similarity data will facilitate more accurate planning and thus enable the manufacturing equipment to be technically utilized in a better manner. The exact definition of individual elements within a working operation permits work measurement to establish smaller time modules, and enables more accurate standard times to be obtained. Then the process planner can utilize technical equipment in an economic manner, e.g. desk computers and so it would be possible to produce the operational chart in a more rational way.

The described relationships are of great importance for the building up of similarity data and therefore call for the consideration of various machining similarity criteria such as

— Geometry,
— Technology,
— Equipment data,
— Machining aids.

The decision data for staged similarity types can be established on the basis of these major influence factors. The criteria combinations defined for the similarity stages of the design families (main, basic and single types) are taken as the basis for development of the similarity types of the machining families. Supplementary machining criteria are defined for machining family build-up. The following explains the three similarity stages:

Similarity Stage III (main type)

The geometric criteria 'basic shape', 'element' and 'arrangement of elements', which have been taken over from the design families, constitute the starting point for the

101

machining families. Further workpiece data such as the 'initial form', 'geometric accuracy' (e.g. 'dimension', 'position-' and 'shape-accuracy') also have to be considered.

As technological criteria, data concerning the 'cutting speed' and 'surface quality' etc. are also of interest. Information concerning the 'power of the machine tools' and 'feed' ect., can also be relevant criteria for the 'equipment data'. Further criteria concerning the machining aids such as 'special equipment', 'clamping tools' and 'cutting tools' etc. also play an important role.

Similarity Stage II (basic type)

In the case of the basic type, the criteria are restricted to the basic shape and elements. The arrangement of elements however, remains optional.

Analogous with the design families, there is also a reduction in the criteria for the operational chart data as compared to stage III of the main types. Depending on their relevance, the criteria could be integrated as a recommencation in stage II or excluded from the consideration altogether. Concerning the geometric criteria, the initial form, for example, is considered with a relevant degree of conclusiveness, whereas the statement in regard to the geometric accuracy is confined to the determination of a general dimensional tolerance.

In establishing the technological requirements, it appears expedient to consider only the 'surface quality' for stage II. In doing so, this statement is combined with the geometric criterion of the general dimensional tolerance.
In view of the selected degree of detailing required for the basic type, the statements concerning the equipment data, e.g. 'machine power' and 'speed' prove to be unnecessary.

The criteria in respect of the machining aids, e.g. 'special equipment' and 'clamping tools', are significant factors in manufacture. From these, we can form clamping tool groups for the individual basic types.

Similarity Stage I (single type)

For the single type, the starting position is confined to established elements only. As a consequence of the low frequency of the single type, it is advisable to keep the operational chart data within strict limits, so that the planning effort and the resultant rationalization effect are kept in a reasonable ratio.

The 'initial form' and a statement concerning the 'representative technical basic shape', which includes machining criteria such as the type and numbers of 'clamping periods' and the internal and external machining, pass for relevant geometric data. This representative basic shape describes an abstract part shape to which the similar part to be planned can be allocated.

The 'special equipment' and 'general tools' may be regarded as being the criteria for the equipment. By considering such criteria, it is possible to reach the first standardization objective, which is limited to selected elements, but leaves the basic shape and the arrangement of elements free. Similarity stage I constitutes the basis of the standardization efforts of stages II and III.

7.2 Procedure: Similarity Types of the Machining Family

In order to attain a rational procedure in the development of similarity types for process planning, a procedure plan is laid down which provides for the following work phases (Fig. 7.4):

— Selection,
— Analysis,
— Synthesis,
— Working aids.

7.2.1 Selection

The selection, in order to establish a limited parts spectrum for a machining family, can be represented by three steps (Fig. 7.4). In the first step, a product group is selected within the production programme. This selection is effected with the aid of the

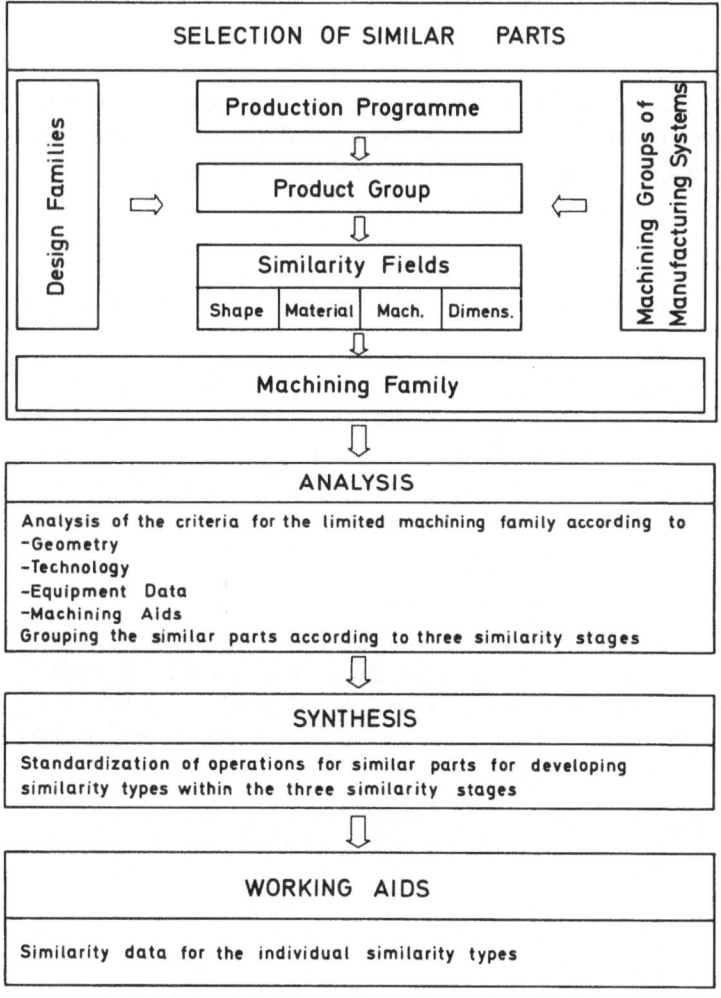

Fig. 7.4. Procedure Plan: Similarity Types of the Machining Family

experience potential of the competent specialist and based on technical-economical considerations.

In a further step, the respective similar parts are selected on the basis of the classification systems for 'technological structure planning'. This happens with the aid of defined similarity fields in respect of shape, material, machining and dimensions.

In establishing the similarity fields, the defined design families and the machining groups of the manufacturing systems are used as a starting point. If sorted parts from the production programme appear which have strong deviations in respect of one or another criteria of the similarity fields as opposed to the entirety, it is advisable to omit them from the allocation in order to get a representative average for the machining family to be investigated.

Investigations have shown that a number of design families can be grouped together in a larger unit to form a starting point for the build-up of machining families.

The last step provides for limitation of the machining family. In doing so, all the similar parts resulting from the design families taken into consideration, are included.

7.2.2 Analysis

Whereas the overall machining stood in the forefront in the selection phase, the consideration in this phase is concentrated on the individual working operations (e.g. turning or milling etc.). The parts can be investigated according to various main criteria. These are the

— Geometry,
— Technology,
— Equipment data,
— Machining aids.

These main criteria, however, can be structured into sub-criteria (Fig. 7.5). The sub-criteria are allocated to the similarity stages according to their relevance. This allocation is influenced by the frequency and also the work shop conditioning factors, e.g. the existing personnel structure and the degree of automation of the equipment. The similar parts are allocated to the individual similarity stages on this basis.

7.2.3 Synthesis

The parts allocated to a similarity stage in the analysis constitute the starting position for grouping the similar parts into similarity types. Those similar parts which unite several of the common criteria are grouped together into a similarity type. It is necessary that all the criteria of these similarity types are maintained. The entirety of the criteria characterizes the similarity type.

Standardized clamping tools and cutting tools can be formed together in groups to provide the necessary machining aids for the machining of all the parts of a respective similarity type.

7.2.4 Working Aids

The realization using similarity planning as a basis requires various similarity data, which are built up depending on the degree of similarity. As can be seen from the pro-

Main Criteria	Sub-Criteria	Similarity Stages		
		Single Types	Basic Types	Main Types
Geometry	Elements	▨	▨	▨
	Basic Shape		▨	▨
	Arrangement of Elements			▨
	Accuracy to Shape			▨
Technology	Surface Quality		▨	▨
	Cutting Speed			▨
	Depth of Cut			
	Tool Life			▨
Equipment Data	Speed Range			▨
	Feed			
Machining Aids	Special Equipment	▨	▨	▨
	Clamping Tools		▨	▨
	Dim. Area of Clamp. Tools		▨	▨
	Supplementation for Clamp. Tools			▨

Fig. 7.5. Allocation of Sub-Criteria to the Similarity Stages of the Machining Family

cedure plan for similarity planning (Fig. 7.2), data concerning the machining family is used on the level of complete machining to describe a workpiece.

In the case of a limited machining family, information is provided in respect of the appertaining product groups and part types as well as the limitation concerning shape material, machining and dimension.

The crucial point is constituted by the working aids which are applied on the level of the working operation. These are

— Summary sheets showing the similarity types,
— Standardized operational data,
— Manufacturing systems and equipment data,
— Cutting data.

The summary sheets help the process planner to allocate an actual workpiece on the basis of its shape to a respective similarity type. The summary sheets are structured in different forms according to the similarity stage for the main, basic and single types. This working aid already provides information in respect of special procedures in the case of variable batch sizes or of existing NC-programming family data.

105

The standardized operational data show the machining conditions for the individual similarity types and are also variably detailed depending on the degree of similarity. The data is valid to established dimensional areas. In the case of the main type, an exact process description of the working operation is provided, including the whole range from the designation of the clamping tool to the establishment of the cutting data. The basic type only contains data concerning cutting tools, clamping tools etc., and leaves the exact operational process indeterminate. The data for the single type provides only general information as to necessary special equipments for a machine tool, general tool, guidelines etc. The data concerning all the degrees of similarity indicate all optimum manufacturing systems and equipment which are possible. Furthermore, they also show alternatives.

The data in respect of the manufacturing systems are harmonized with those of the operations, and include all the information concerning the possibilities of individual machine tools within a manufacturing system. In this way the workpiece requirements for a similarity type can be compared with the capabilities of the existing manufacturing systems and allocated to the optimum manufacturing system.

The cutting data provides general working information and contains the interrelationship between cutting depths, tool life, cutting speed etc., which is shown in the form of diagrams. In view of its specific character, this data is only suitable for describing the main types.

7.3 Procedure: Similarity Types of the NC-Programming Family

Similar to the case of the machining families, the build-up of data for NC-programming families is based on a three-stage flexible planning system. The aim here is to simplify the effort needed for the preparation of programmes for NC-machining in respect of quality and quantity and, at the same time, to limit the planning variety. Thereby it also results in a reduction of the variety of cutting tools required for manufacture. The procedure considers the frequency analysis made for the build-up of machining families. The geometric characteristics 'basic shape', 'elements' and 'arrangement of elements' are used as a basis. Technical criteria in respect of programming constitute the focal point of grouping of parts for NC-programming families. The analysis has shown that the NC-programming families should vary within surveyable limits in order to allow a simple handling of the similarity data by the programmer. The standardized programmes resulting from the similarity data should be machine-independent and describe the workpiece requirements on the machine.

The same procedure technique as used for the machining families can be applied to transmit the rules of GT to NC-programming.

7.4 Model Example: Similarity Types of the Machining Family

Selection

A parts spectrum of the product area 'pumps' was selected for the model example. This example represents a continuation of the design family 'wear rings' mentioned in Chapter 6, and builds up on the same basis.

In conceiving the design family, only the shape and material fields were required for preliminary sorting with the aid of similarity fields. In the case of the machining families, the machining fields — as a limitation of the complete machining area — and the dimensional fields — as a limitation of the machine working area — are included for the selection process.

In this example, the machining and dimensional fields for the machining family 'F' are (Fig. 7.6):

— Machining fields (code explanation see also Fig. 4.27):
  Code 01:   Turning,
  Code 02:   Turning and grinding,
  Code 03:   Turning, grinding, milling and bench work,
  Code 09:   Turning, drilling and bench work.

— Dimensional fields (code explanation see also Fig. 4.28):
  Code CH: $D = 100 - 250$ mm $\Big\}$ for rotary machining
             $L = 20 - 100$ mm

             $A = 100 - 250$ mm $\Big\}$ for facing and drilling machining
             $B = 100 - 400$ mm

Machining Family F			
Product Group		DA	High-pressure centr. pumps
		DB	Low-pressure centr. pumps
		DC	Bore-hole pumps
Type of Part			Wear ring
Similarity Fields	Shape	00	Cutting processes, rotary parts without deviation, disc-shaped and flanges without thread
	Material	0 / 4 / 5	Steel without heat treatment / Casting, Iron without Alloying Elements / Alloy Cast Iron, Malleable Iron
	Machining	01 / 02 / 03 / 09	Turning / Turning and Grinding / Turning, Milling, Grinding and Bench Work / Turning, Drilling and Bench Work
	Dimension	CH	D = 100-250 mm / L = 20-100 mm $\}$ for Turning Work / A = 100-250 mm / B = 100-400 mm $\}$ for Facing and Drilling Work

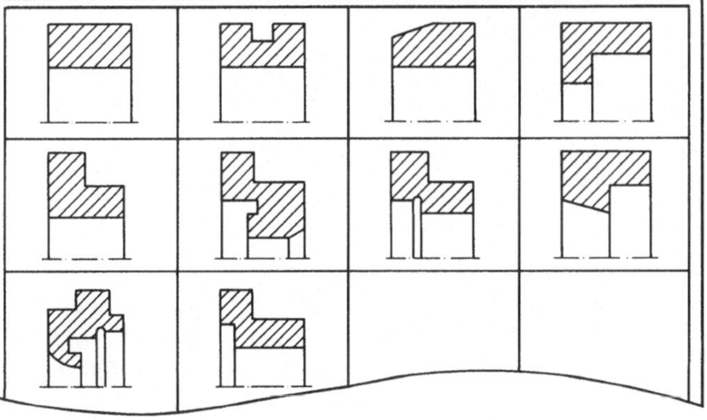

Fig. 7.6. Delimitation of Machining Family F

The described similarity fields represent a coarse sieve for the parts selection and limits in this case the machining family F. The machining fields indicate the possible working operations for the machining family F. The sequence of working operations is not considered in the selection phase. In order to prevent a too extensive sub-division of the machining family in the coarse phase, the sequence of working operations is determined only in the phase of detailing. The limitation of the similarity fields was also accomplished in considering the machining groups of the GT-manufacturing systems.

Analysis

In the working phase of the analysis, the machining criteria for the individual working operation are established on the basis of the similar parts of the limited machining family F. In this case, 'turning' is used as an example. Concerning this analysis, the main criteria 'geometry', 'technology', 'equipment data', 'machining aids' and have been selected (Fig. 7.5).

The existing personnel structure and the degree of automation of the equipment should be included when establishing the degree of similarity. The sub-criteria for the similarity stages are allocated on this basis. This means the required degree of similarity of the similarity stages is fixed.

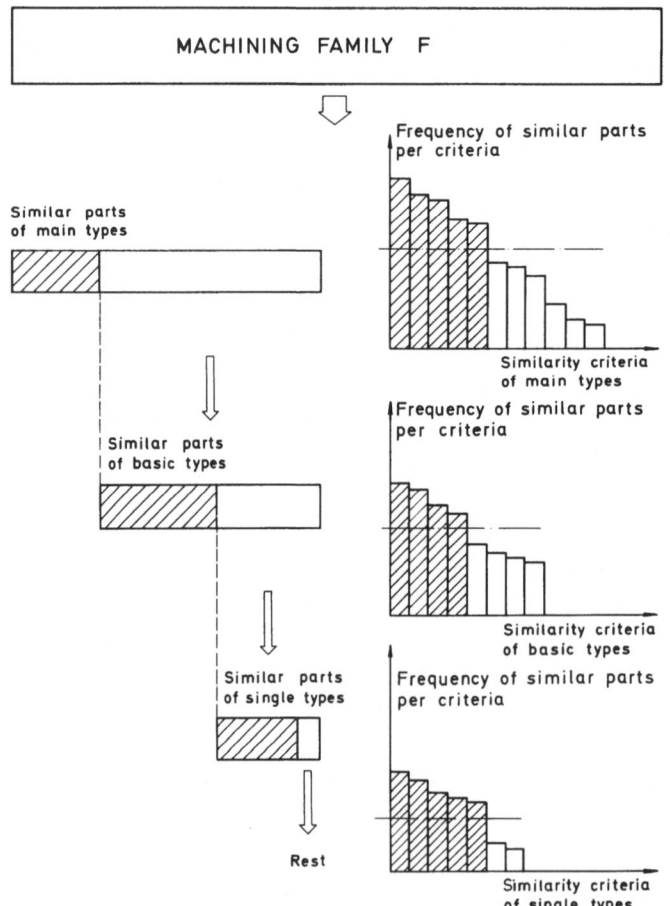

Fig. 7.7. Analysis –
Allocation of Similar
Parts to Similarity
Stages

108

In a further step, the similar parts are allocated to the similarity stages according to their frequency. The selection is concentrated first to the similar stage with the highest degree of similarity, that means to the main types (Fig. 7.7). Those main type criteria which comply with a minimum number of similar parts are designated for the main type synthesis. The remaining similar parts are then investigated in respect of a reduced number of criteria in the lower similarity stage of the basic types. If a minimum number of similar parts can be allocated to the individual criteria then these parts are attached for grouping to a basic type. The remaining parts are then investigated according to the criteria for the single types. If, during this consideration, certain criteria fall short of a minimum number of allocated similar parts, then they are omitted from the consideration within the framework of similarity planning.

Synthesis

Fig. 7.8 shows as an example, the necessary steps to be taken in the case of the synthesis for the basic type B. The criteria which were found out by the analysis for the basic types and the similar parts allocated to this similarity stage constitute the starting position.

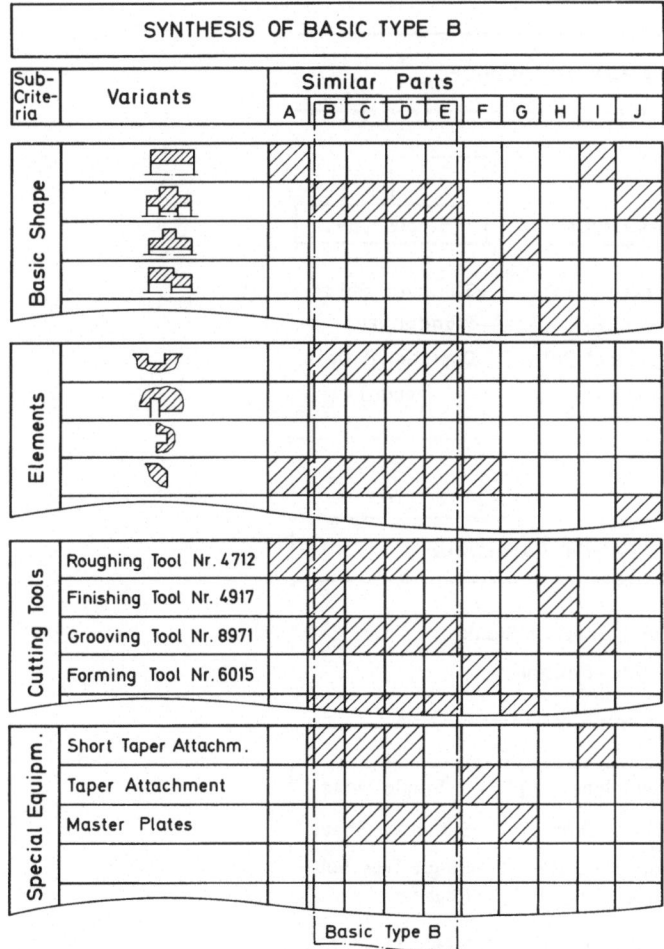

Fig. 7.8. Synthesis – Grouping the Basic Type B

Building up on the analysis, the similar parts B, C, D and E — which have several common criteria — are grouped together in the synthesis within the basic types to form the basic type B. The stated cutting tools provide a standardized cutting tool group. The clamping tools group also presents itself in the form of various required chucks. The synthesis for the basic type B is effected on the basis of the inclusion of all the criteria belonging to the basic types as they are represented in Fig. 7.5

Working Aids

Fig. 7.9 provides a summary of the necessary aids within the framework of the machining families and the NC-programming families. These are structured according to the three similarity stages and coordinated with various similarity types.

As the selection of the NC-programming families is made from the machining families the data of both systems should be coordinated with each other. They also are based on the single, basic and main types.

The summary sheets for the similarity types show the examined parts on the basis of their defined characteristics: basic shape, elements and arrangement of elements.

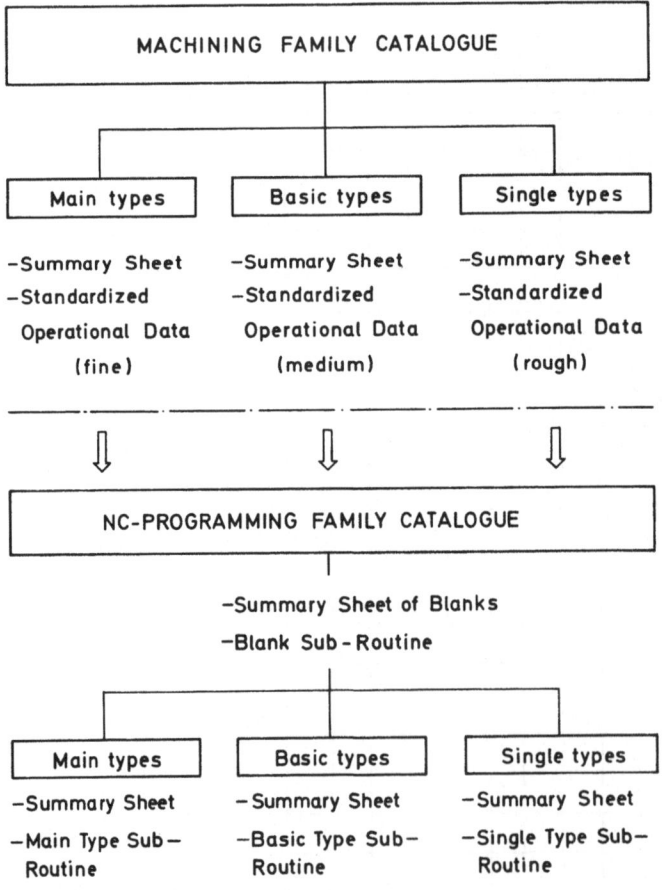

Fig. 7.9. Summary and Interrelationship of the Working Aids for Machining and NC-Programming Families

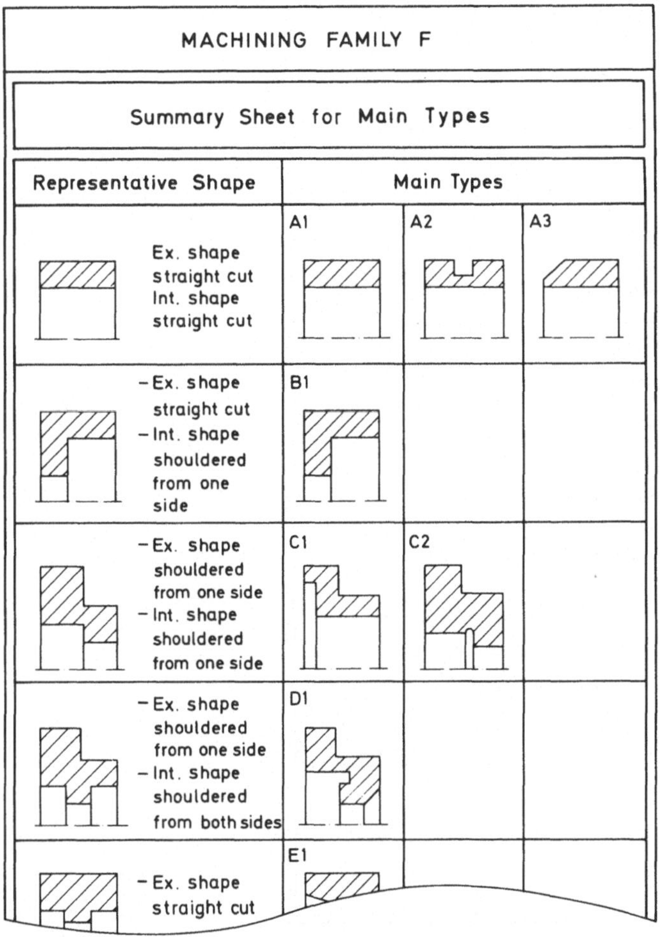

Fig. 7.10. Working Aids – Summary Sheet of the Main Types

Fig. 7.10 provides a summary of the main types within the framework of the machining family F. On the occasion of the occurence of a planning event, the process planner can find whether the part to be planned may be allocated to an existing main type on the basis of the shape characteristics. If this is not the case, it is to be classified on the basis of the data belonging to the next lower similarity stage of the basic types, if a corresponding similarity type is available. In the objective concept, a similarity type with the highest degree of similarity i.e. a main type, is always aimed at.

If a similarity type can be found on the basis of its defined characteristics, the machining conditions or the procedure to be followed within this working operation can be ascertained from the standardized operational data. The data refers to the established main dimensional areas for the individual similarity types. The conclusiveness rises with the increasing degree of similarity and describes in advance the working operation more accurately.

Whereas, in the case of the single types, the stated characteristics are restricted to general information (Fig. 7.11), which shows the necessary equipment structure in a gener-

111

MACHINING FAMILY F		
Standardized Operational Data for Single Type A		

Representative Shapes

External and Internal
Machining

Characteristics	Variants	
Initial Form	Sawn Casting	
Process	Engine Lathe Work	
General Tools	External Turning Tools Internal Turning Tools Internal Grooving Tools – Back Facing Tools – Forming Tools	
Special Equipment	Short Taper Attachment	

GT - Manufacturing Systems		
Optimum	Alternative	NC - Programme
GT – Flow Line GT – Cell 201.1 GT – Centre 101.1 101.4	101.3	P – 1107

Fig. 7.11. Working Aids – Standardized Operational Data of Single
Type A

al form (e.g. machining internal or external, information concerning the initial form, process, general cutting tools and special equipment), additional requirements are placed on the machining aids with the basic type (clamping tools equipment and cutting tool groups as well as data concerning the geometric and technological accuracy) (Fig. 7.12).

The data of the main type (Fig. 7.13) however, describe the partial operations within the operation in detail, and indicate the necessary clamping tools, cutting tools, accuracy to size, to shape and of position, special equipment, equipment data such as the machine power, speed range feed and also technological data like cutting speed, surface quality, material etc. All the standardized operational data contains information concerning the optimum and alternative manufacturing systems. Data showing the existing standardized NC-sub-routines provides the junction point to the NC-programming families. The cutting data represents a general working aid and serves as a supplementation to the standardized operational data. Because of their specific conclusive character, however, they are only used to establish the cutting conditions for the main type.

The technically optimum allocation of a similarity type to a GT-manufacturing system presumes an exact knowledge of the workpiece requirements with regard to the machine tools on the one hand and the capabilities of the machine tools on the other hand. The workpiece requirements for the machine tools are shown in the standardized operational data. The machine capabilities are to be established with the aid of working operation matrices in order to adapt the similarity data to the latest state of existing manufacturing systems [33] (Fig. 7.14). The matrices, however, only represent an indirect working aid for the process planner and are determined for each type of process as 'turning without threading' or 'milling'. They are ascertained as a result of the allocation of the process-specific machining characteristics.

The requirement in respect of the parts spectrum, the capabilities of the machine tool in question, and thus the manufacturing system are established on the basis of this matrix. This matrix has been structured in such a way that the workpiece requirments of a main, basic or single type can be used for machine selection. The determined optimum manufacturing system is shown in the standardized operational data for information purposes.

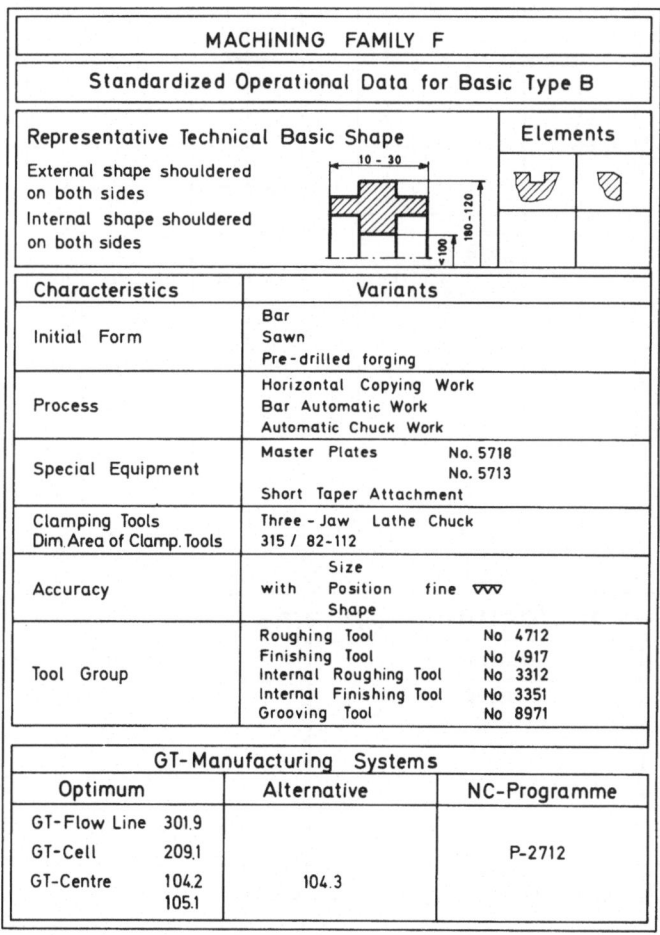

Fig. 7.12. Working Aids — Standardized Operational Data of Basic Type B

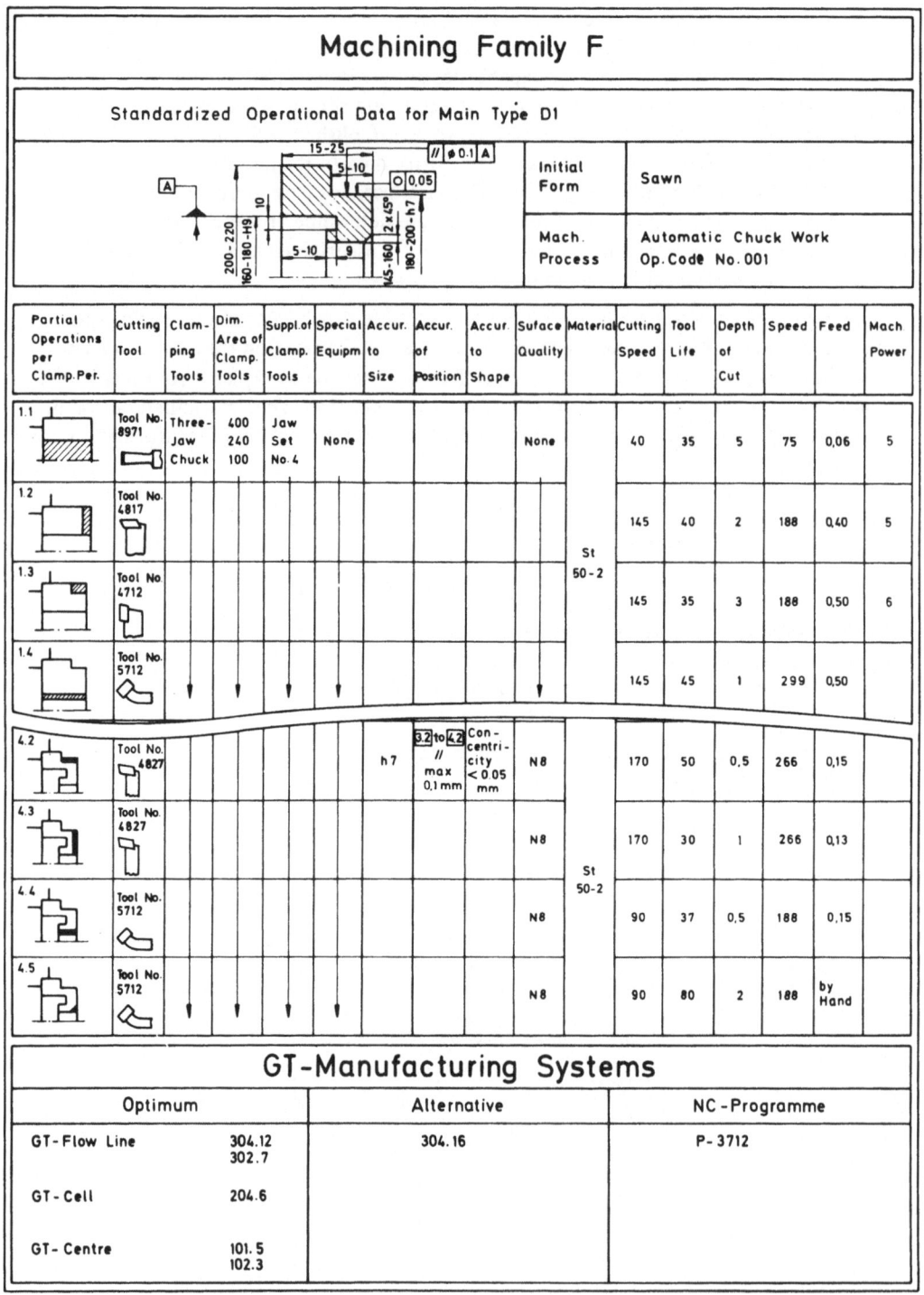

Machining Family F

Standardized Operational Data for Main Type D1

Initial Form	Sawn
Mach. Process	Automatic Chuck Work Op. Code No. 001

Partial Operations per Clamp. Per.	Cutting Tool	Clamping Tools	Dim. Area of Clamp. Tools	Suppl. of Clamp. Tools	Special Equipm	Accur. to Size	Accur. of Position	Accur. to Shape	Suface Quality	Material	Cutting Speed	Tool Life	Depth of Cut	Speed	Feed	Mach Power
1.1	Tool No. 8971	Three-Jaw Chuck	400 240 100	Jaw Set No. 4	None				None		40	35	5	75	0,06	5
1.2	Tool No. 4817										145	40	2	188	0,40	5
1.3	Tool No. 4712									St 50-2	145	35	3	188	0,50	6
1.4	Tool No. 5712										145	45	1	299	0,50	
4.2	Tool No. 4827					h 7	0.2 to 4.2 // max 0,1 mm	Con-centri-city < 0.05 mm	N 8		170	50	0,5	266	0,15	
4.3	Tool No. 4827								N 8		170	30	1	266	0,13	
4.4	Tool No. 5712								N 8	St 50-2	90	37	0,5	188	0,15	
4.5	Tool No. 5712								N 8		90	80	2	188	by Hand	

GT-Manufacturing Systems

Optimum		Alternative	NC-Programme
GT-Flow Line	304.12 302.7	304.16	P-3712
GT-Cell	204.6		
GT-Centre	101.5 102.3		

Fig. 7.13. Working Aids – Standardized Operational Data of Main Type D1

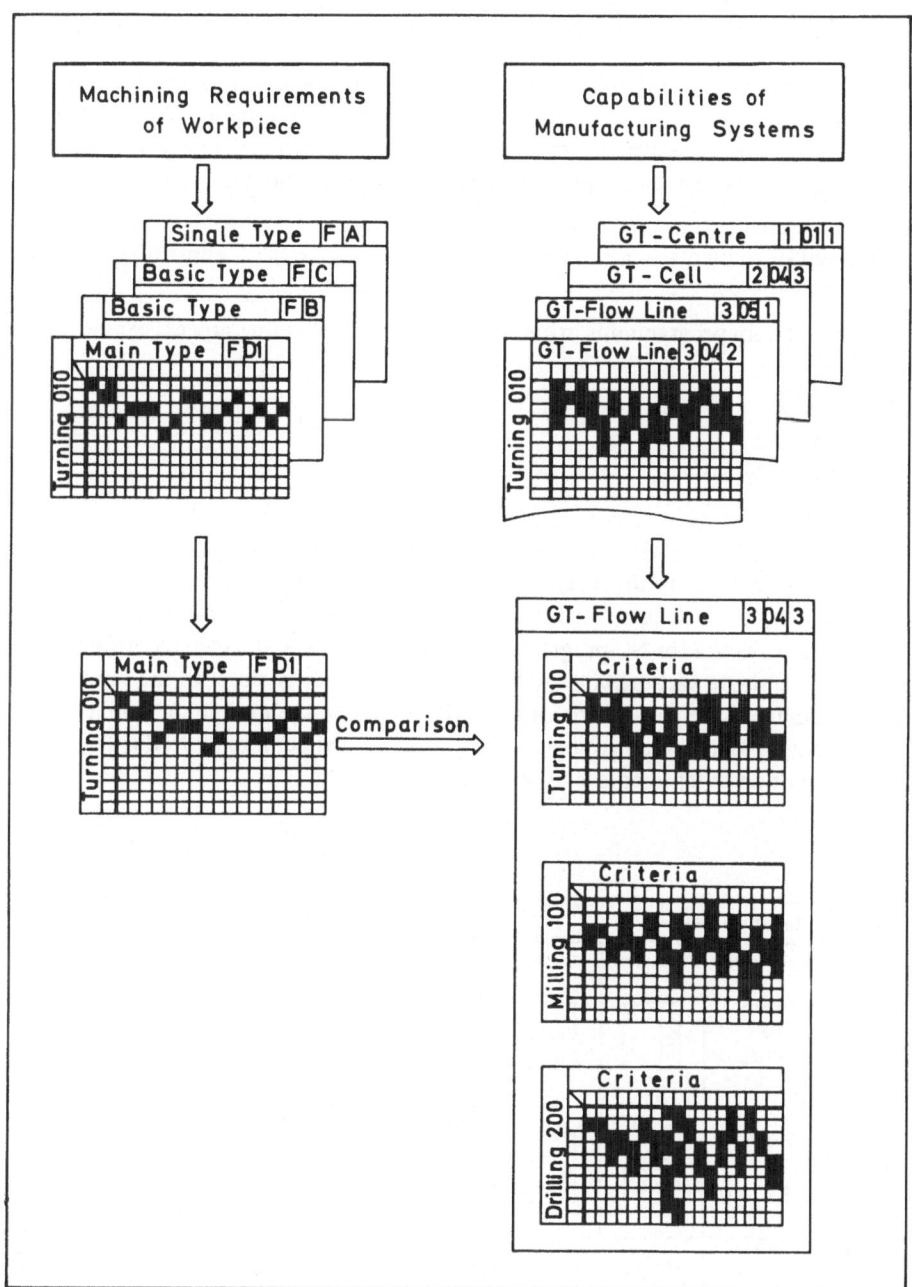

Fig. 7.14. Operation Matrices for GT-Manufacturing Systems

7.5 Model Example: Similarity Types of the NC-Programming Family

In the planning phase 'selection', the parts of the machining family are singled out which, on the basis of technical and economic considerations, appear to be suitable for NC-machining. Consequently, workpieces were selected which can be worked with the same chuck and do not require special equipment on the machine.

The criteria combinations of the geometric characteristic 'basic shape', 'elements' and 'arrangement of elements' – as a function of the similarity stage – constitute the basis for the build-up.

In addition, certain programming criteria concerning the planning process can be derived (Fig. 7.15), i.e.

– Cutting tool,
– Blank contour,
– Surface quality,
– Clamping tool,
– Machine power.

In an analogous manner to the machining families, the criteria are now laid down according to their conclusiveness for the three different similarity types. For the synthesis, the specific criteria for NC-programming of the individual parts are identified on the basis of the single part drawings within the same similarity stage. In the same grouping manner as the elaboration of similarity types for the machining families, the NC-programming families are grouped according to several common criteria. The resultant grouping of parts represents a NC-programming family.

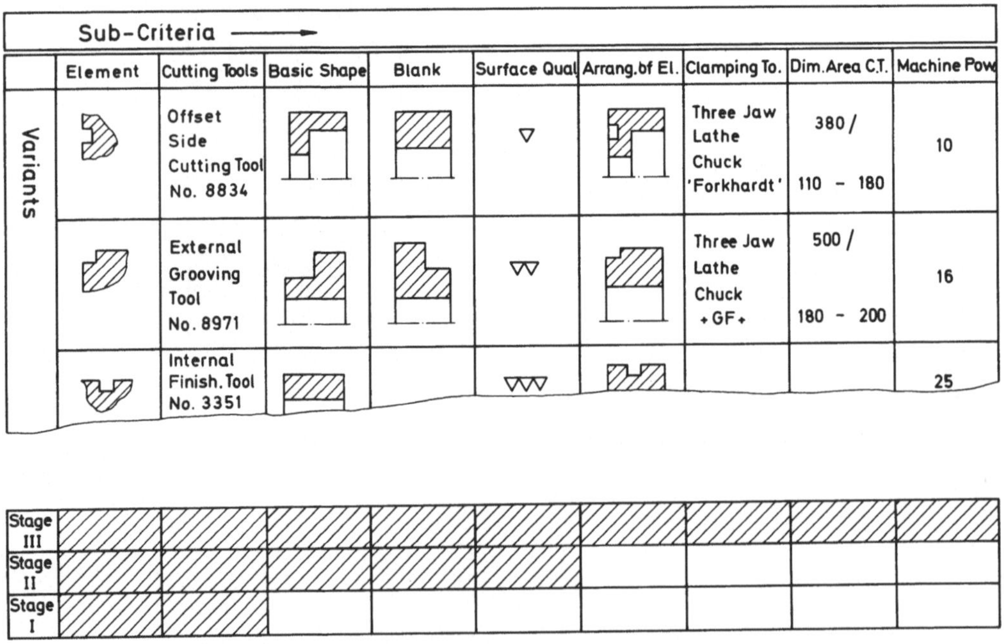

Fig. 7.15. Allocation of the Sub-Criteria to the Similarity Stages of NC-Programming Family

In the sense of the different functional stages, various working data results in the planning phase 'working aids' (Fig. 7.9): This data concerns

— Summary sheets of main, basic and single types,
— Summary sheets of blanks,
— Blank sub-routines with blank-sketch,
— Main type sub-routines,
— Basic type sub-routines,
— Single type sub-routines.

The summary sheets of main, basic and single types are structured according to the similarity stages (Fig. 7.16). In this summary sheet, the limited parts spectrum of the individual similarity stages is shown in the form of a summary. For the similarity stage III, the main types are shown with regard to the basic shape, elements and arrangement

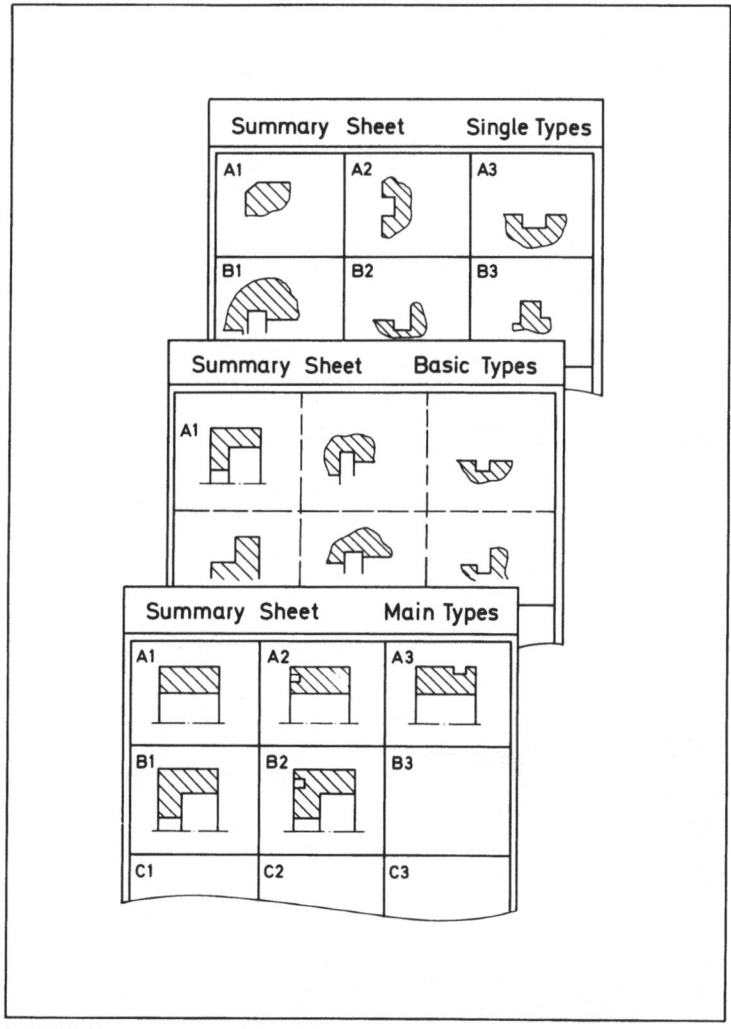

Fig. 7.16. Summary of the Similarity Types for the three Stages of a NC-Programming Family

117

of elements. The basic types of similarity stage II are represented in a summary sheet which shows the similarity types in the basic shape and the appertaining elements. In the case of similarity stage I, the representation is limited to the machining elements with the aid of the summary sheets of single types.

The summary sheets of blanks (Fig. 7.17) show the existing and usable blank contours for the limited parts spectrum of the wear rings for NC-machining. The blank contour thereby can be given by means of non-cutting processing (e.g. forging, pressing or casting) or cutting processing (e.g. rough turning). A blank sub-routine with corresponding blank sketch is built up for every listed blank contour. At the occurrence of a planning event, the planner can add the effective actual dimensions to the blank sub-routine and the blank sketch.

The blank description is designed in the same way for every blank, i.e. independent of the similarity stages. As a function of the similarity stages, the NC-programming data contains various amounts of information.

The 'main-type sub-routine' (Fig. 7.18) includes a sketch of the finished part with alphabetical-numerical dimensions. The sub-routine is given in the form of a standardized NC-programme for the finished part.

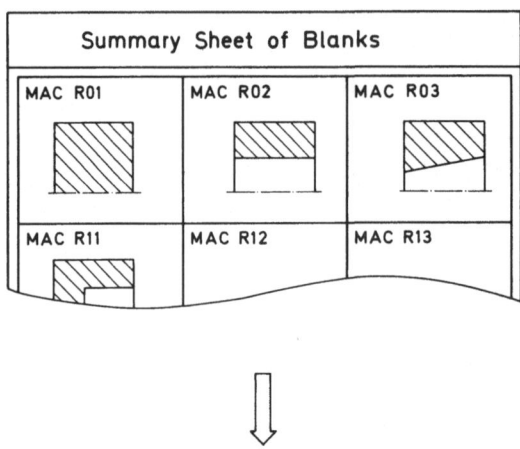

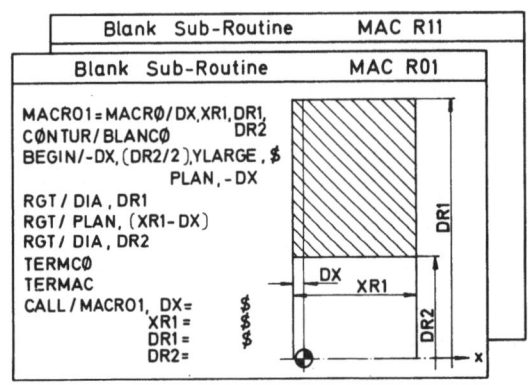

Fig. 7.17. Summary of the Blanks and Blank Sub-Routines within a NC-Programming Family

118

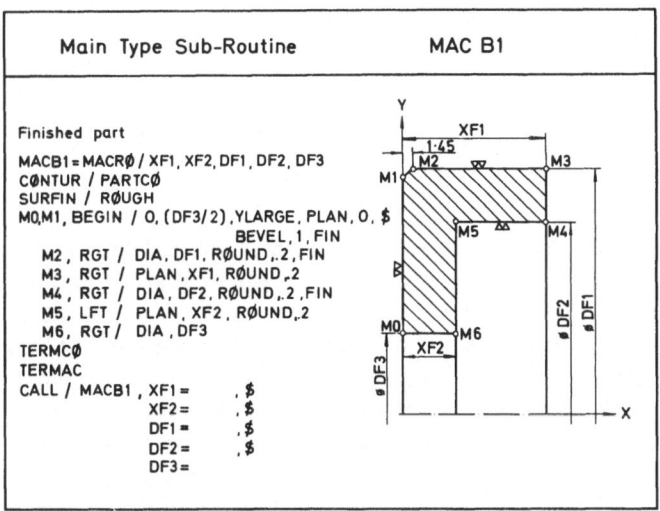

Fig. 7.18. Main Type Sub-Routine for NC-Programming Families

The basic type sub-routine includes several similar parts, which are also established in a composite finished part sketch. All the corresponding programme instructions are included in the basic type sub-routine.

Due to the lesser frequency of the single type for similarity stage I, only the 'programme segments' are noted for the stated elements in the similarity data. This programme segment contains the programme instructions for the element to be machined. The dimensions are shown in an alphabetical-numerical order in the sketch and based upon a fictitious zero point.

7.6 Aspects of Computer Aided Process Planning

The similarity principle, which is described in the proposed solution, is characterized by a short, medium and long-term objective.

The short-term objective tries to prepare simple drawings and operational chart indexes with a minimum effort. These working aids provide improved access to already completed similar planning activities and thus allows a certain rationalization effect to be attained with this data storage.

The medium-term objective provides for the creation of data for similarity types as described in this work. This already results in appreciable rationalization effect in respect of a reduction of the process planning work, an improved accuracy of the planning work and a standardization of the machining aids.

The long-term objective is aimed at the creation of an economic basis for the introduction of computer assisted process planning. The standardization process includes in the main the limitation of the types variety for process planning and manufacture. This has consequences for the attachments, the clamping tools and also the cutting tools. In this way, it is possible to develop tool lists and setting plans which can be printed out with the aid of the computer.

8 Work Measurement

Work measurement plays a particularly important role in the planning and development of GT, because the manufacturing time represents the basis for the various fields of application, e.g. cost calculation, production control, wage structure etc. To begin with, the work measurement techniques mentioned in the literary references [43, 44, 45] are briefly presented. The investigations have shown that a three-stage procedure on the basis of the similarity planning is also recommended for work measurement. In doing so, the three similarity types (main, basic and single types) again stand in the foreground. The build-up of work measurement data is described with the aid of a model example and the possibility of determining standard time with the aid of the computer is also discussed.

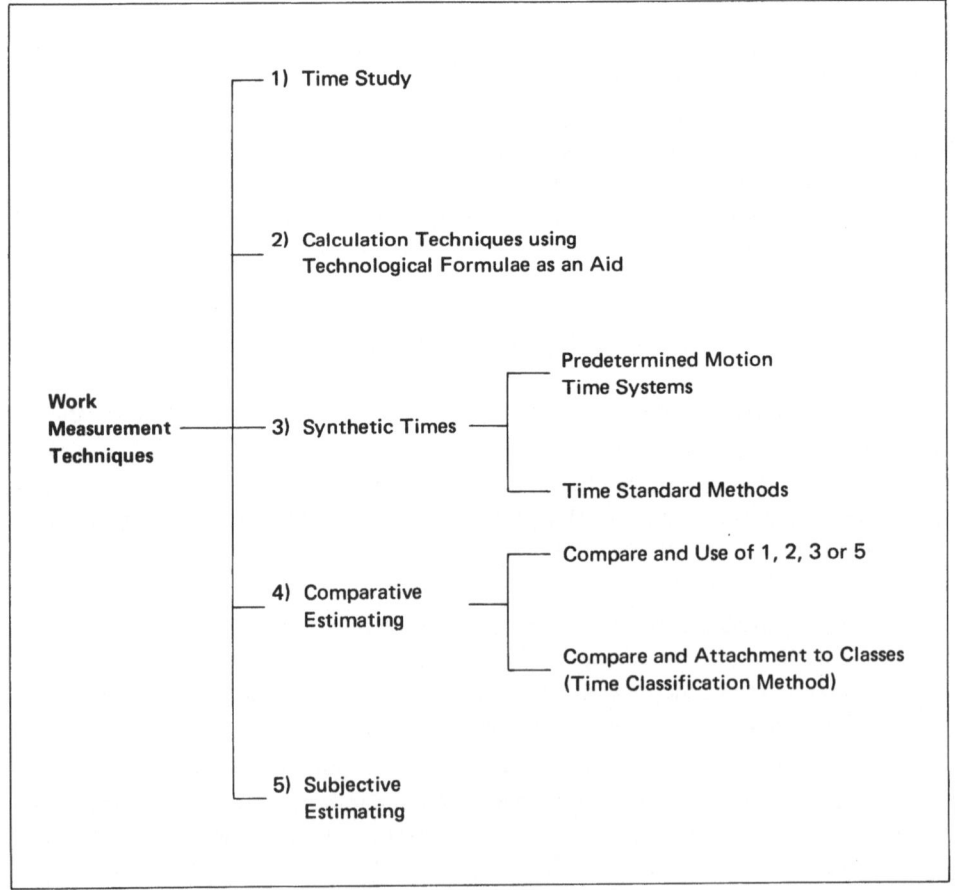

Fig. 8.1. Work Measurement Techniques

8.1 Survey of Work Measurement Techniques

According to a corresponding proposal of Wojda work measurement can be effected by means of the following five techniques (Fig. 8.1) [46]:

— Time studies (recording and evaluation of actual times),
— Calculation techniques using technological formulae as an aid,
— Synthetic times,
— Comparative estimating,
— Subjective estimating.

8.1.1 Time Studies

In this case, the observation techniques

— Stop-watch measurement,
— Activity sampling

stand principally in the forefront for the determination of the effective times.

'Stop-watch measurement' can be made by a time study specialist or with time recording equipment. In the case of working operations that are effected over a longer period, the time spent can be recorded by the operator himself.

'Activity sampling' — as an observation technique — is applied for multiactivity studies. Multiactivity recording consists of registering the frequency of previously determined activities for a single or several machines with the aid of short-time observations carried out at random.

8.1.2 Calculation Technique Using Technological Formulae as an Aid

With the aid of technological formulae, the time is calculated for certain parts of a working operation, particularly for those carried out by machines, e.g.

$$t = \frac{L}{n \cdot s},$$

time t (min), length L (mm), speed n (rev/min), feed s (mm/rev).

8.1.3 Synthetic Times

The synthetic times are based on the synthesis of time modules for defined work contents, which are then grouped together for a certain working operation. The predetermined motion time systems contain time modules for human motions.

When analysing working operations, we continually encounter movement sequences which repeat themselves with varying degrees of frequency. They can be the general and conventional activities existing in numerous fields, or special activities that only occur over a limited area. This has led to the development of the method of time standards for limited, definite application areas. The following will briefly explain the two types of synthetic times used in the measurement work.

Predetermined Motion Time Systems

The systems of predetermined motion time are the smallest time modules for basic human motions such as reaching, grasping, moving and fitting. Two such well known systems are work factor and methods time measurement (MTM) [44, 47].

In the case of work measurement, the MTM method is perhaps more widely known because it is a closed system with various concentration stages. The three generally applied MTM-systems − MTM-1, MTM-2 and MTM-3, with various degrees of concentration, complement each other and can be combined according to the purpose of application.

Generally applied MTM-systems can hardly extend beyond the concentration stage attained with MTM-3 without losing extensively in accuracy.

If, in the future, work measurement is to be effected still more economically with MTM, then certain developments will have to be taken. These will apply to a particular field of application and therefore the tendency will be to develop MTM-systems which enable the times to be measured more quickly and with an acceptable degree of accuracy, within the specific fields of small batch production.

One of these area orientated data systems has been recently developed by the Swedish MTM-Association in cooperation with reputable Scandinavian industrial and advisory firms. In this case, a concentration on the machine tool area, giving special attention to one-off and small batch production, has been effected. The new system is known as MTM-V (V being the first letter of the Swedish word 'verktygsmakiner = machine tools tools). MTM-V consists of 12 different work elements, which are matched to the machine tool work (e.g. operating, setting, checking, etc.). The aspired high speed of application is attained principally through the development of larger work elements. By way of statistical calculations, each individual time for a work element is assured in such a way that the attainable accuracy of the system remains within reasonable limits.

A further advantage to be gained with MTM-V, as also with other MTM-systems, is that it can be applied for the direct analysis of a working activity as well as for the build-up of calculation sheets. In favourable circumstances, only one fifth to one tenth of the time needed for the working operation is required for the analysis. In its accuracy and speed of application, MTM-V is exactly defined and can therefore be combined with MTM-1, MTM-2 and MTM-3 without difficulty [47].

Time Standard Methods

The determination of time standards is particularly important for activity sectors which always reoccur. The size of the sector for the time standard value is dependent on its intended application. The advantage of time standard values is found in their time-saving application. They can consist of time modules for the pre-determined motion time systems, as well as time studies. If quantitative influence factors with several measurement values are experienced for a certain activity sector, then the time standard values can also be determined with the aid of correlation and regression calculations. The time standard methods constitute a good fundamental basis for the build-up of time calculation data in the field of small batch production.

8.1.4 Comparative Estimating

The fundamental principle of comparative estimating is to determine by comparing the machining tasks of that working operation which is very similar to the operation to be planned. If a difference exists between the operation obtained by means of the comparison, for which a standard-time is already available, and the activity to be calculated, the time is determined by the aid of addition and subtraction of corresponding elements. The following procedures can be employed for building up comparative estimating:

— Time studies,
— Calculation techniques using technological formulae as an aid,
— Synthetic times,
— Time classification method.

The 'time classification method' can be regarded as comparative estimating because it is based on similar machining tasks and the standard time is determined by way of comparison. Instead of a single value, however, as for example when calculating predetermined motion time systems, a classification value is applied. The time classes can be laid down, for example, in the form of a geometrical series. With the time classification method, however, no additional times are evaluated for the extra or deleted elements of a working operation. On the contrary, it must be decided whether the deviation entails an allocation to another class than that of the comparison workpiece.

The fundamental condition for the application of comparative estimating is the existence of similar machining tasks as it is met within the framework of GT. With the aid of comparison criteria — in the form of similarity data — the similarity can be determined and thus a time comparison is made possible.

8.1.5 Subjective Estimating

This estimating technique provides for the approximate determination of standard times from memory and experience. In the case of more comprehensive working operations, it is advisable to structure the work content into partial operations and to make the estimation individually to ensure a reduction in the overall error factor.

8.2 Work Measurement on the Basis of the Similarity Types

The similarity stages of process planning constitute the basis for the build-up of work measurement data. Apart from the similarity characteristic and the recurring frequency the allocation of individual cases to the respective stage is dependent on a number of other marginal conditions, e.g. in the case of machines with high investment and operating costs where an exact work measurement is important for the utilization of the manufacturing capacity and for supervising the economics. Here, a higher degree of detailing would have to be selected than that corresponding to the work measurement data.

The degree of detailing has a direct influence on the accuracy of the work measurement and the respective time effort. The emphasis, however, should not be placed on the accuracy of the individual case but on the accuracy that is obtained from the total of individual cases within a certain period of time, e.g. a month. Within this concept, the statistical consideration constitutes the basis for work measurement, and is related to the wage structure which is dealt with in Chapter 10.

The main objective is to build up work measurement data in the following manner: Reduction of the effort required for work measurement with a simultaneous increase in the respective accuracy by using similarity types as a basis.

The selection of the degree of detailing is dependent on the expected frequency of the work measurement data. The studies show quite clearly that even in this GT application area the required flexibility with regard to the most economic elaboration of the work measurement data results from the three-stage method on the basis of the similarity types (main, basic and single type). The allocation of the individual case, as well as the selection of the degree of detailing is effected by a specialist in the process planning department. On the other hand, the degree of detailing for the work measurement data — in the form of comparative estimating or special calculation sheets for the individual similarity types of the machining families — are obtained from past similarity and frequency analysis. Two problems have to be faced in the building up of work measurement data for the three similarity types, namely the choice of techniques for determining and evaluating the data, and the form and structure of the working equipment for the user. The establishment of influential factors for the machining families and the allocated GT-manufacturing system with the relevant machines are a pre-condition for the collection and evaluation of data.

The proposed techniques for the individual similarity types and the resultant working equipment are briefly explained.

Main Type

Because of the high degree of similarity of the allocated parts spectrum, the predetermined motion time systems are recommended for the elaboration of the work measurement data for the main type with a high degree of detailing at the work elements stage. The results of the investigated firm show that the MTM-2 method is particularly suitable for manufacturing areas with medium-size batches and a high recurrence frequency. In the textile machine area at the investigated firm, 80% of work measurement is now effected on the basis of MTM-2. In the small-batch production field with a wide workpiece spectrum the studies showed that MTM-V can be used in most cases for work measurement. In the case of recurring human motions for the operation of machine tools, MTM-2 with its increase accuracy is recommended. Further help may be found in the data catalogues, which have been elaborated by institutes and advisory firms. These catalogues contain information on various application areas, e.g. transport and storage, measurement, inspection, erection, surface treatment, cutting etc. The data can be purchased and adapted to the local workshop conditions.

Basic Type

For the basic type, we can use the same techniques as are applied for the main type. In contrast to the main type, however, larger work sectors are laid down and these are characterized by larger inaccuracy. Individual time tables are drawn up for the particular work elements of the basic type for the purpose of increasing the degree of accuracy of the work measurement.

Single Type

With the single type, it is not expedient to make a detailed build-up of the work measurement data because the frequency of application is too low. The collection of empirical values, which are entered on comparison sheets and established according to the relevant dimensional criteria and influence factors, may be used for this similarity stage. Work measurement is effected by comparison and estimation.

The time classification method can also be applied for the single type provided that sufficient empirical data is available which can be supplemented by time studies. For the evaluation, it is advisable to use correlation and regression calculation to determine the compensation functions between the individual empirical values. In order to be able to establish a mathematical formula for work measurement, the overall cycle time for the individual working operations and the relevant influence factors have to be determined when using this technique. The resultant time standard values can be shown in the form of tables or diagrams.

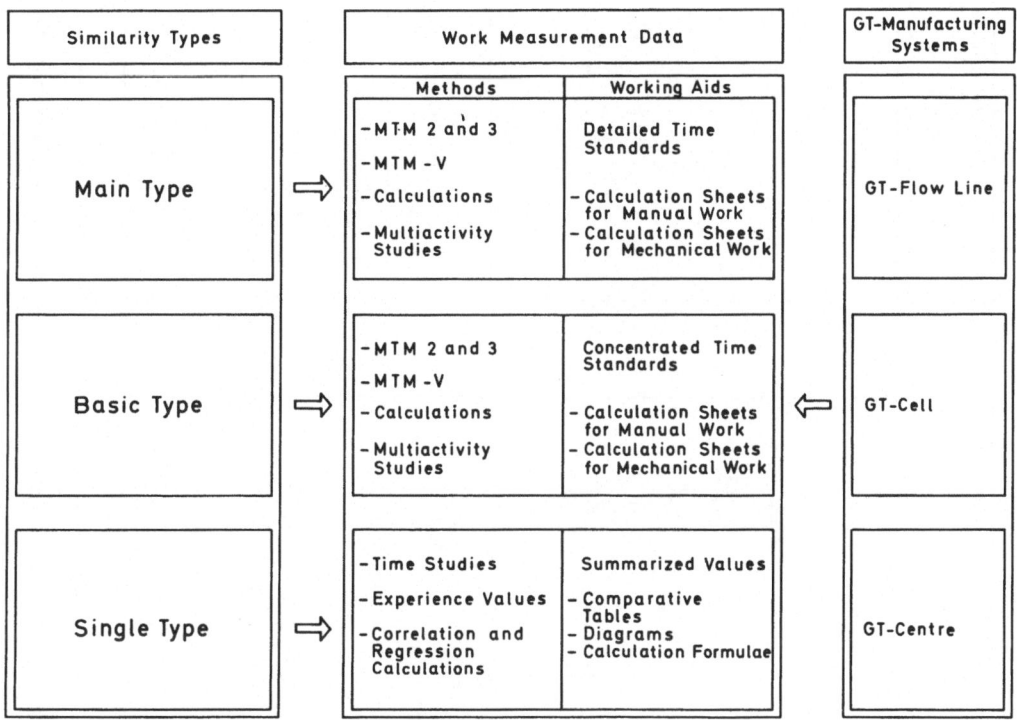

Fig. 8.2. Techniques and Working Aids of Similarity Types

Fig. 8.2 summarizes again these statements, whereby the recommended work measurement techniques and working aids are allocated to the individual similarity tpyes giving due consideration to the similar machines of GT-manufacturing systems.

8.3 Model Examples

For work measurement on the basis of similarity types, some possible solutions are presented with the aid of a model example for wear rings (Fig. 8.3). This proposed solution leaves enough latitude for other forms of representation, which can always be adapted to meet the requirements of the respective factory organization.

125

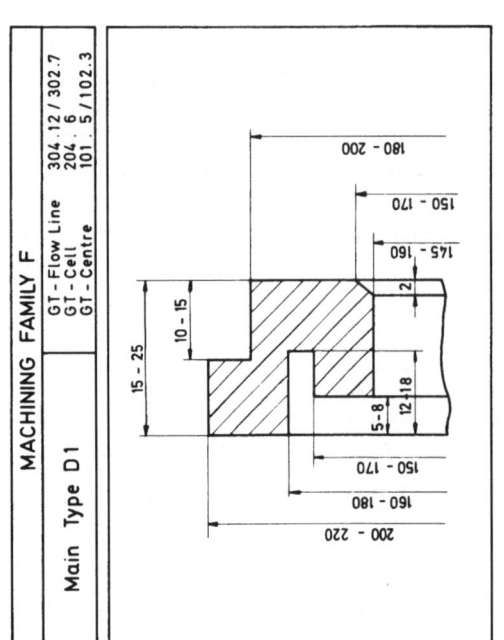

MACHINING FAMILY F		
	GT - Flow Line	304.12 / 302.7
	GT - Cell	204. 6
	GT - Centre	101. 5 /102.3
Main Type D1		

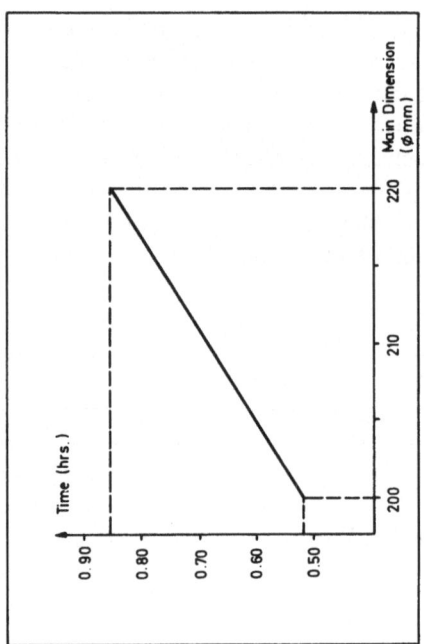

Fig. 8.4. Working Aids for Determination of the Standard Time

Wear – Ring

Fig. 8.3. Example of a Wear Ring

4 Bore Holes

8.3.1 Main Type

The work measurement data (for the main type) consists of standardized sheets which are structured according to the various categories of time elements, e.g. machine-controlled, clamping, outside times etc.

As a result of the high similarity with the main type, in most cases the standard time can be determined with the aid of a simple diagram. The external diameter represents the reference value in this example (Fig. 8.4). The other dimensions can be neglected because they are related to a certain ratio to the external diameter. For the diagram build-up detailed calculations have to be made for the various reference values. Fig. 8.5 and 8.6 show the standardized data for machine-controlled times, clamping and outside times within the framework of the main type and the allocated manufacturing systems. The specified 'time values' have been determined with the aid of the MTM-V method and technological calculation formulae. The setting-up times and the summarization of the standard time are shown as standard values in Fig. 8.7.

8.3.2 Basic Type

As opposed to the main type, for which the standard time can be determined with the aid of a simple diagram, the work measurement for the basic type has to be determined in different steps because of the larger variety (Fig. 8.8). A nomogram which duly considers five variables has therefore been developed for determining the machine-controlled time (Fig. 8.9). The outside times are calculated with the aid of standard values as in the case of the main type. Fig. 8.10 shows these values for the basic types and also for the machine-controlled time determined by means of the nomogram (Fig. 8.9).

8.3.3 Single Type

The well known comparison sheet is recommended as the form of representation for the single types. It contains the various individual single type variants and the relevant influencing factors. Fig. 8.11 shows an example of a comparison sheet for the single type. With a sufficient number of empirical values, this comparison sheet can take the form of a time classification diagram or be supplemented with the aid of a mathematical formula of work measurement on the basis of regression and correlation calculation.

8.4 Prospect of Computer-assisted Work Measurement

In view of the large quantity of data needed for work measurement the question arises how far a computer can be used as an aid. According to present-day knowledge there appears to be two application possibilities, namely on the one hand for the build-up of work measurement data, and on the other hand for the direct work measurement in the individual case. Preliminary experiences [48] show that there is a rationalization possibility for the practical application.

MACHINING FAMILY F

Main Type D1

GT - Flow Line 304.12/302.7
GT - Cell 204.6
GT - Centre 101.5/102.3

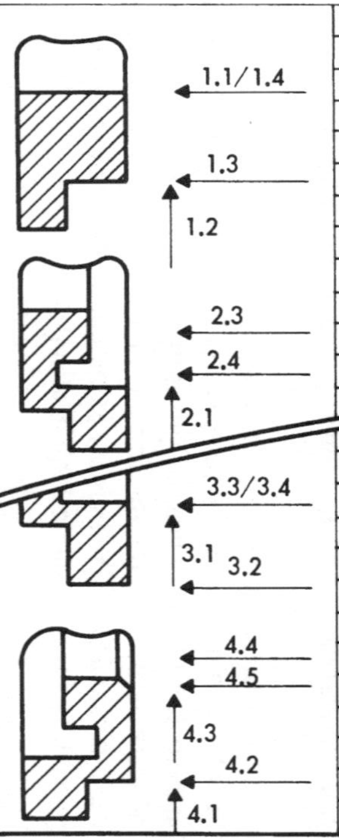

Standard Values: Partial Operations per Clamping Period	Clamping Tool	Saddle v	Saddle h	Machining ∇	Machining ∇∇	Machining ∇∇∇	Pl.T.	Cyl.T.	Tool Nr.	Varying Values: Blank ø	Turn ø	Cutt. Speed	s	f	Depth a	Number of Cuts	Lenght l	Lenght b + e	Total Lenght L	L x i	Tabel Value per 10 mm	Machine-Contr. Time in TMU
1.1	A1/3-Jaw Ch.	X		X				X	8971		148	40	75	0.06	5	1	26	4	30	30	3530	10'590
1.2			X	X			X		4817	240	149	145	188	0.40	2	2	46	4	50	100	220	2'200
1.3		X		X				X	4712	240	191	145	188	0.50	3	10	12	2	14	140	180	2'520
1.4		X		X				X	5712	148	149	145	299	0.50	1	1	22	4	26	26	110	286
																				296		15'596
2.1	A2/3-Jaw Ch.		X	X			X		4817	240	149	145	188	0.40	2	1	56	4	60	60	220	1'320
2.2		X		X				X	4712	240	222	145	188	0.50	3	6	9	4	13	78	180	1'404
2.3		X		X				X	8961		160	40	75	0.06	4.5	2	5	2	7	14	3530	4'942
2.4		X		X				X	8961		170	40	75	0.06	4.5	1	15	2	17	17	3530	6'001
									8961											169		13'667
3.4		X			X			X	8961	162	160	165	316									
3.5		X			X		X		5722	160	150	165	316	0.05	1	1						
3.6		X			X			X	5712		45°	165	316	b.H.	1	3	Chamfering				110	330
																				81		5'984
4.1	A3/3-Jaw Ch.		X		X		X		4827	220	190	170	266	0.13	1	1	15	2	17	17	500	850
4.2	Soft Jaws		X		X	X		X	4827	191	190	170	266	0.15	0.5	3	13	2	15	45	400	1'800
4.3		X			X		X		4827	190	149	170	266	0.13	1	1	21	4	25	25	500	1'250
4.4		X			X			X	5712	149	150	90	188	0.15	0.5	2	15	4	19	38	560	2'128
4.5		X			X			X	5712	150	154	90	188	b.H.	2	1	2	2	4	4	1770	708
4.6		X			X			X	5712		45°	90	188	b.H.	1	3	Chamfering				110	380
																				129		7'066

Fig. 8.5. Calculation of the Machine-Controlled Time

MACHINING FAMILY F

Main Type D1	GT - Flow Line 304.12 / 302.7
	GT - Cell 204.6
	GT Centre 101.5 / 102.3

Work Element	Time Value in TMU	Partial Op. per Clamping Period							
		1		2		3		4	
		No.	Time Val.	No.	Time Val.	No.	Time Val.	No.	Time Val.
3 - Jaw Chuck	920	1	920	1	920		920		920
Setting by Eye	2430	1	2430	1	2430				
Setting with Watch	3240						3240		3240
Cleaning with Air	300								300
Cleaning of Chips per Clamping	540	1	540	1	540		540		540
Allowance for Clamp. and Taking off per 5 kg	20	1	20						
Chip Protection Shield Close / Open	130	1	130	1	130				
Outside Time (Clamping) in TMU			4040		4020		4700		5000

Work Element	Time Value in TMU	Partial Op. per Clamping Period							
		1		2		3		4	
		No.	Time Val.	No.	Time Val.	No.	Time Val.	No.	Time Val.
Switching Machine on/off Changing r.p.m.	70	4	280	3	210	3	210	3	210
Feed on/off	30	15	450	10	300	5	150	7	210
Changing the Feed	120	3	360	3	360	4	480	4	480
Changing the Tool Holder	270	3	810	3	810	2	540	2	540
Saddle Clamping / Loosing	80	1	80	1	80	2	160	2	160
Cylindrical Turning, Facing, Grooving, Chamfering	90	15	1350	10	900	5	450	8	720
Move Tailstock Back and Fore	210								
Measuring	100	4	400	4	400	6	600	8	800
Outside Time in TMU			3730		3060		2590		3120

Fig. 8.6. Standardized Outside Time

MACHINING FAMILY F			
Main Type D1	GT - Flow Line	304.12 / 302.7	
	GT - Cell	204. 6	
	GT - Centre	101. 5 / 102.3	

Type of Time	Partial Op. per Clamping Period			
	1	2	3	4
Outside Time (Clamping)	4040	4020	4750	5000
Outside Time	3730	3060	2590	3120
Total Outside Time	7770	7080	7290	8120
Machine - Controlled Time	15596	13667	5984	7066
Cycle Time per Operation	23366	20747	13274	15186
Contingency Allowance per Operation	3505	3112	1991	2278
Standard Time in TMU	26871	23859	15265	17464
Standard Time in hrs. per Clamping	0,27	0,24	0,16	0,18
Standard Time in hrs.	0,85			

Work Element		Time Value in hrs.	Partial Op. per Clamping Period							
			1		2		3		4	
			No.	Time Val.	No.	Time Val.	No.	Time Val.	No.	Time Val.
Basic Setting- up Time	Simple Parts	0.12	1	0,12	1	0,12				
	Difficult Parts	0. 25					1	0,25	1	0,25
Use of Soft Jaws	Without Radial or Axial Run out Tolerances	0.10					1	0,10		
	With Radial or Axial Run out Tolerances	0.20							1	0,20
	Use of 4 Jaw Chuck	0.17								
	Between Centers	0.17								
Clamp. Attach- ment	Clamping with Screws on Face Plate	0.24								
	Fastening with Taper	0.15								
	Steady Rest	0.15								
	Use of Behind Saddle	0.07	1	0,07	1	0,07	1	0,07	1	0,07
	per Cutting Tool	0.05	3	0,15	3	0,15	2	0,10	2	0,10
	per Tolerance	0.02					2	0,04	1	0,02
	Setting-up Time without Cont. Allow. in hrs.			0,34		0,34		0,56		0,64
	Setting-up Time with Cont. Allow. per Clamp in hrs.			0,40		0,40		0,65		0,75
	Setting-up Time in hrs.		2.20							

Fig. 8.7. Standard and Setting-up Time

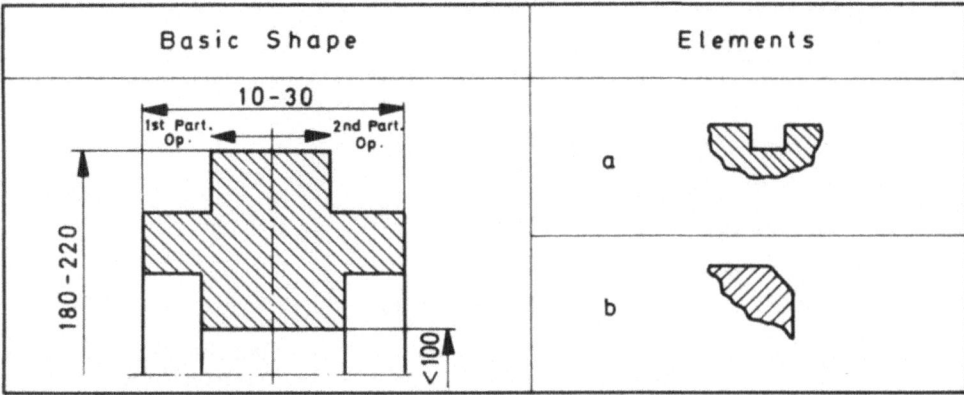

	MACHINING FAMILY F	
Basic Type	GT - Flow Line GT - Cell GT - Centre	304.12 / 302.7 204. 6 101 . 5 / 102.3

	Basic Shape	Elements	
	10-30 1st Part. Op. 2nd Part. Op. 180-220 <100	a	
		b	

Type of Time	Partial Operation	
	1	2
Outside Time (Basic Shape + Element a and b)	12.7	10.9
Machine-Controlled Time (Basic Shape + Element a and b)	16.3	13.9
Cycle Time per Operation	29.0	24.8
Cycle Time per Operation plus Contingency Allowance	33.3	28.5
Standard Time in hrs. per Partial Operation	0.55	0.45
Standard Time in hrs.	1.00	

Work Element		Time Value in hrs.	Partial Operation			
			1		2	
			No.	Time Val.	No.	Time Val.
Basic Set-ting-up Time	Simple Parts	0.12	1	0.25	1	0.25
	Difficult Parts	0.25				
Use of Soft Jaws	Without Radial or Axial Run out Tolerances	0.10				
	With Radial or Axial Run out Tolerances	0.20			1	0.20
	Use of 4 Jaw Chuck	0.17				
	Between Centers	0. 17				
Champ. Attachment	Clamping with Screws on Face Plate	0.24				
	Fastening with Taper	0.15				
	Steady Rest	0.15				
	Use of Behind Saddle	0.07				
	per Cutting Tool	0.05	6	0.30	3	0.15
	per Tolerance	0.02	2	0.04	2	0.04
	Setting-up Time without Contingency Allow.			0.59		0.64
	Setting-up Time with Contingency Allow.			0.68		0.74
	Setting-up Time in hrs. per Part. Op.		0.70		0.75	
	Setting-up Time in hrs.		1.45			

Fig. 8.8. Working Aids for Determination of Standard and Setting-up Time

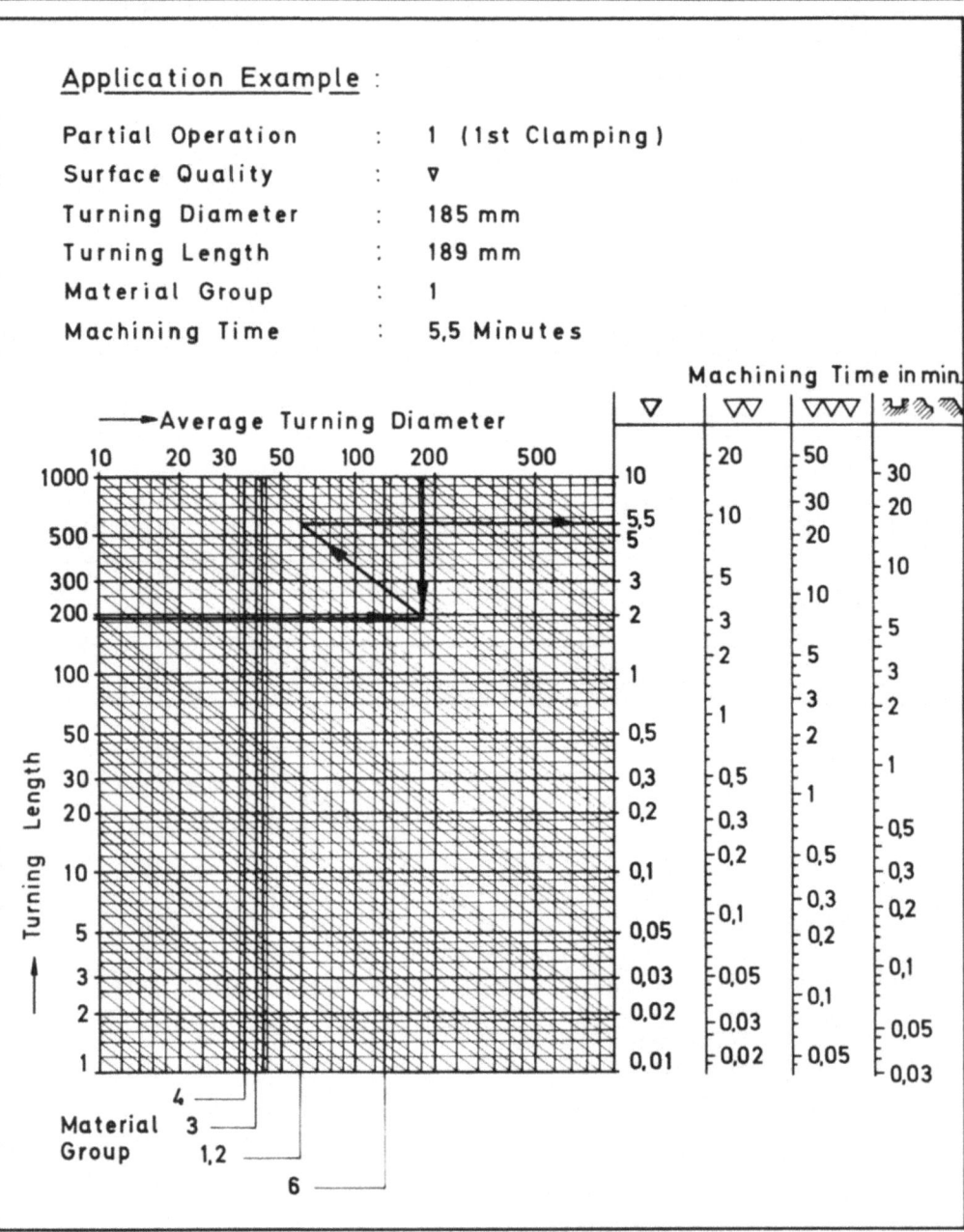

Fig. 8.9. Example for Determination of Machine-Controlled Time

132

MACHINING FAMILY F

Single Type A	GT - Flow Line	304.12 / 302.7
	GT - Cell	204.6
	GT - Centre	101.5 / 102.3

Operation

Cylindrical Turning, Shouldering, Facing, Internal Turning with Groove and Bore Hole

Sketsch

Workpiece 1   Workpiece 2

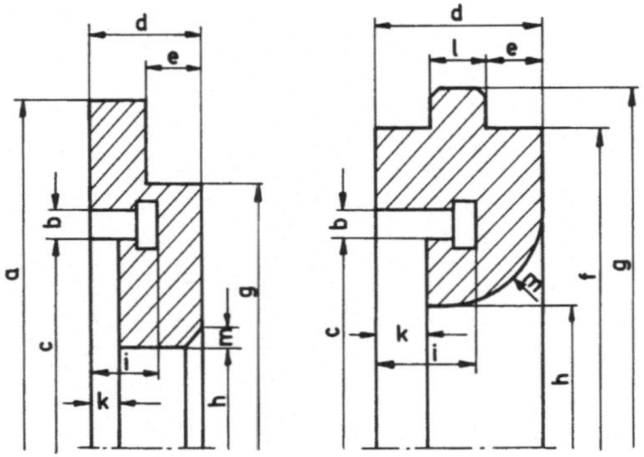

Article	Material	Sketch Dimensions										Operation			Remarks
Drawing No.	Workpiece	a	b	c	d	e	f					Ts	Tst		and Date
		g	h	i	k	l	m								
4 - 052.435	CK 45	220	10	160	20	12,5						1.00	1.50		
"	1	180	150	15	5		45°								
3 - 052.789	CK 45			10	215	67	10	230				2.40	2.00		
"	2	250	190	50	30	10	R34								

Fig. 8.11. Comparison Sheet

MACHINING FAMILY F

Basic Type B	GT - Flow Line	304.12 / 302.7
	GT - Cell	204.6
	GT - Centre	101.5 / 102.3

Surface Quality and Elements

		Partial Operation			
		1		2	
		#	Time Value in min	#	Time Value in min
		L		L	
▽		185	5,5	185	5,2
		189		170	
▽▽		185	3,8	185	4,2
		71		77	
▽▽▽				190	4,2
				30	
a		160	7,0		
		72			
b				155	0,3
				4	
Machine - Controlled Time in min			16,3		13,9

Fig. 8.10. Machine-Controlled Time and Standardized Outside Time

Allowance per	Time Value in min	Partial Operation			
		1		2	
		Number	Time Value in min	Number	Time Value in min
Basic Type	5,9	1	5,9	1	5,9
Clamping	1,4	2	2,8	2	2,8
Tolerance	0,8	2	1,6	2	1,6
a	0,8	3	2,4		
b	0,6			1	0,6
Outside Time in min			12,7		10,9

133

9 Production Control

Production control in manufacturing and assembly is built up on the fundamentals of process planning, which comprises all the technical and organizational measures, e.g. the establishment of working operations and their sequence, the optimum allocation of manufacturing systems and equipment as well as the programming, if numerically controlled machines are applied. Production control, however, concerns the measures required to execute an order in the sense of process planning. This phase is of great importance especially when we consider, that about 7000 different parts or positions are needed for example when an order is received for a marine diesel engine. As a result of this, approximately 16500 deadlines are involved, i.e. 16500 activities must be executed to the required moment if the finished product is to leave the works on the agreed date. This example shows quite clearly the amount of data that has to be processed. In addition, a large number and different customer orders are also being processed at the same time, so that the number of activities and deadlines to be considered increases in an immense manner.

This consideration shows that if we can realize GT-lines and GT-cells to an increasing degree in the manufacturing area during the layout planning phase, it will enable us to simplify the production control work appreciably because only the beginning and completion of the manufacturing order would have to be controlled and no longer the individual working operations. This also follows from the work carried out by the Institute for Science and Technology at the University of Manchester which shows quantitatively that production efficiency can be improved even more with the application of GT-manufacturing systems [49]. This simplification also applies to the combining of similar manufacturing orders and their sequence within the framework of GT-lines and GT-cells. In opposition to functional layouts, where the machine tools with similar operations are grouped together, the combining of similar manufacturing orders to so-called 'additive series' places stringent requirements on the production control and data processing. In the event of modifications or break-downs, it becomes difficult to handle because of the additional resultant influence factors. This statement is made clear by Fig. 9.1, where the planning of orders for an additive series is shown. The individual workpiece must be coordinated with regard to the deadlines and the location aspect for each of the specified group operations 10, 20, 30 etc. [50]. In practice, there is a restriction to individual working operations. This also applies to the GT-centres in this concept.

9.1 The Crucial Points of Application for GT

With production control, the critical points of the influence exercised by GT lie principally in the elaboration of the work programme and the work distribution. In the case of the elaboration of the work programme, it entails grouping together the manu-

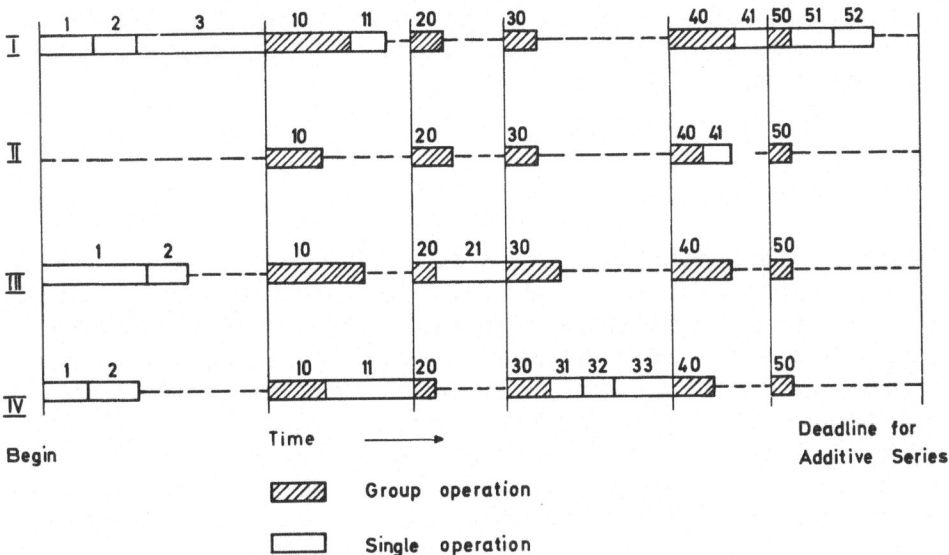

Fig. 9.1. Planning of Manufacturing Orders Belonging to the same Additive Series

facturing orders from the previous coarse loading for every planning period in such a way that optimum conditions are created in respect of deadline, loading and throughput. With work distribution, however, it concerns the optimum sequencing of the manufacturing orders and preparing the required work data, material and working aids in accordance with the work programme [51].

On the one hand, the planning of the manufacturing orders within a planning period is accomplished according to the urgency of the deadlines, and on the other hand according to the technological sequence aspect. The work programme contains the manufacturing orders that will have to be effected for a certain manufacturing system, e.g. GT-cells, in the subsequent planning period. In accordance with the product structure, the planning period can be of one, two or three weeks duration. Depending on the duration of the planning period, there is a possibility of grouping similar manufacturing orders in series with the purpose of aiming for the best possible total throughput time. So-called 'sequence-families', based on the similarity criteria of the workpiece machining, which influence the sequence, are used as a working aid here. The primary purpose of the sequence family is to support the optimum technological sequence planning for manufacturing orders within a planning period through the selection of similarity criteria. An additional task for the sequence families is the determination of alternative variants if the manufacturing system should be overloaded. The general aim of the sequence families application consists in raising productivity and reduce throughput times.

The following partial objectives stand therefore in the foreground:

— Reduction of the effort involved in the preparation of tools,
— Reduction of setting times,
— Reduction of training times,
— Harmonization of the work rhythm,
— Indication of possible alternative variants in the event of overloading.

For practical applications, it is expedient if the similarity criteria of the sequence families are available in the form of classification numbers which can be sorted according to the arbitrary aspects of sequence planning.

9.2 Sequence Family Criteria

The task is now to convert the machining technique laid down in the operational chart into a suitable form for the sequence planning of the manufacturing orders. In selecting the criteria, care had been taken that they can also be used for functional layouts, so that flexibility can be achieved between the individual manufacturing system forms, which is positively effective especially in the case of alternative variants. To ensure a simple handling and survey, the criteria are selected in a process-neutral way. In the preliminary studies it was noted that, apart from several characteristics from the classification system for the workpiece and manufacturing description, further sequence-determining criteria have to be considered. The information of the sequence families should serve as a starting basis for the preparation of the work programme and work distribution. In selecting the criteria, this concept is limited to a few relevant and simple to handle decision factors. This restriction is made possible by the stepwise grouping of workpieces with similar machining requirements, because a form of pregrouping has already been effected through the layout planning, the parts design with the aid of design families and the subsequent process planning on the basis of the machining families. The development of sequence families therefore represents the final link of the GT-families and is used to support the production control within the framework of the GT-manufacturing systems. The influence area is limited by the deadline priority of the order-dependent manufacture, as it is especially the case in mechanical and plant engineering. A general method for the sequence is selected from these marginal conditions as shown below. Automated places of work and manufacturing systems are an exeption for additional savings can be made by, for example, specific sequence planning as it is shown in the work by Meyer [52]. Detailed decision criteria can therefore be derived for these cases and with the aid of matrices, optimum sequences can be determined.

The following will introduce the individual criteria for the building up of general sequence families. The suggested classification numbers are intended to provide an idea as to the characteristics of the individual criteria.

9.2.1 Main Criteria

Basic Forms of the Manufacturing Systems

The basic forms of the GT-manufacturing systems constitute the first criteria of the sequence families. As various forms may occur in practice, the functional layout will be included as well as the GT-manufacturing systems.

With these first criteria for the sequence families, we are able to express at the same time the working operation sequence of the allocated manufacturing system (loose, flexible or fixed). This information has proven to be useful particularly in the case of displacement due to overloading of individual manufacturing systems. The search for alternative variants should be made firstly at the same manufacturing system level, and only if no possibility is found here, can it be turned to the next lower stage. With this approach, the displacement of a manufacturing task can be advantageously carried out from the optimum total throughput standpoint.

The following proposal could be used to classify these criteria:

Code	Basic Forms of Manufacturing System
0	Functional layout
1	GT-centre
2	GT-cell
3	GT-flow line
.	.
.	.
.	.
9	Transfer line

Machining Process, Dimensional Areas

The operational code of the classification system (see Chapter 4), represents the second criteria for the sequence families. This code describes the machining process and dimensional areas for the working operation and for the machine tools allocated to a manufacturing system. The dimensional areas of the workpieces are given as upper and lower marginal limits. Accordingly, this criterion has the following build-up:

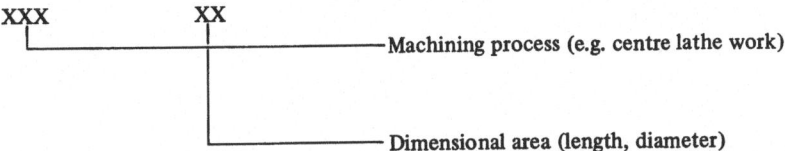

This criterion is included in the operational chart and can be obtained from the basic data for the deadlining. This information becomes increasingly important in particular in the case of alternative variants if manufacturing orders have to be displaced to another manufacturing system because of overloading.

This basis makes possible a coarse classification which, depending on the degree of complexity of the manufacturing order, requires further detail criteria, if a decision is to be made concerning the concrete planning or alternative possibilities.

9.2.2 Additional Criteria

The criteria described in the following section are assisting the decision-making process, but do not replace it. Depending on the kind of process and manufacturing task, so many influencing factors would be required to attain an automatic loading compensation for machine tools that the respective economic use would not be warranted at present time. As a result of the numerous internal and external disturbance factors during the progress in one-off and small batch production, we cannot resort to an examination by an expert.

Material Classes and Initial Forms

By selecting an optimum sequence for the material classes or the initial forms, the setting times for the individual manufacturing orders can be improved in regard to tool and clamping changes.

The material classes and initial forms can be taken from the workpiece supplementary code (see Fig. 4.14) and are specified in the operational chart.

Code	Material Classes
0	Steel, non alloyed and alloyed < 5% without heat treatment
1	Steel, non alloyed and alloyed < 5% with heat treatment
2	Steel, alloyed > 5% without heat treatment
3	Steel, alloyed > 5% with heat treatment
4	Cast iron, no alloying elements
5	Alloy cast iron, malleable iron
6	Cu and Cu-alloys
7	Al and Al-alloys
8	Other materials
9	Several materials

Code	Initial Forms
0	Casting
1	Forging
2	Bar
3	Pipe
4	Ingot
5	Sheet metal ≤ 12 mm
6	Sheet metal > 12 mm
7	Stock parts
8	Other forms
9	Several forms

Loading Time for the Manufacturing Order

As further criterion, the loading time was selected for the batch size. It is determined in the following time classes:

Code	Loading Time
0	≤ 1 hour
1	≤ 2 hours
2	≤ 4 hours
3	≤ 8 hours
4	≤ 16 hours
5	≤ 24 hours
6	≤ 32 hours
7	≤ 40 hours
8	≤ 80 hours
9	> 80 hours

If the existing production control system already has this information in an uncoded form, it can be used as a direct reference and the code number would no longer be required. The criteria 'loading time' for the sequence family should assist the following recommendations:

In preparing the work programme for a certain period of time, care should be taken that in planning the manufacturing orders a certain sequence is achieved in accordance with the duration of order. A too pronounced mixture between short and long-term orders often leads to a situation where the long-term orders have to be interrupted in favour of the short-term ones because of deadline bottlenecks. Furthermore, we have to take into consideration that in the event of a work displacement, the short-term orders are treated first. The reason for this is that in process planning the allocation is made on the basis of the optimum manufacturing preliminary conditions. With alternative variants however, a certain loss is to be expected in productivity. The code number for the loading time provides the work distribution with information regarding the loading situation so that decisions can be made with the aid of the mentioned recommendations.

Special Equipment

The criterion 'special equipment' is intended to provide information concerning the extent and the kind of special tools and equipment necessary for the execution of a manufacturing order. With the following proposal the degree of complexity of the equipment increases according to the increasing code number:

Code	Special Equipment
0	None
1	Special tools
2	Pre-set tools
3	Simple work, measuring and testing equipment
4	Special attachments
5	Feed and receiving attachments
6	Double-deck machines
7	Special testing machines
8	Interlinking machines
9	Special equipment combinations

This criterion should provide the work distribution with a first indication that with an increasing code number a displacement in another manufacturing system involves difficulties and a corresponding reduction in productivity is to be expected.

Job Demands

The job demands, which are corresponding with the manufacturing order to be executed, represent a further criterion. As a displacement can only be made in general with the same machining process, an attempt is made to express the degree of demands by way of simple criteria.

Code	Demands
0	Simple
1	Medium
2	Special
3	Highest

This rough grouping should not be compared with job evaluation as in this case the demands are compared for all the places of work, in opposition to the sequence family where a relative comparison is only made within the same process. This information also serves for the sequence planning and loading compensation.

Similarity Degree of the Parts Spectrum

The criterion in respect of the selected similarity degree of the parts spectrum is similar to the definitions described in Chapter 7.

Code	Similarity Degree of the Parts Spectrum		
0	Single type	low degree of similarity with general work instructions	
1	Basic type	medium degree of similarity with specific work instructions	
2	Main type	high degree of similarity with description of work operations	

With this criterion, it should also be pointed out that at the occasion of the work distribution the higher code numbers will lead to difficulties with the alternative variants. They have a detrimental effect on the work quality, and can only be compensated through additional effort in respect of preparation and inspection.

9.3 Sequence Family Structure

As shown in the previous description, the sequence family is principally an aid for the technological sequence planning in the drawing up of work programmes and the preparation of alternative variants. The criteria sequence of the sequence family has been selected on the basis of the most frequent decision cases. In this concept, they are the

— Basic form of the manufacturing systems,
— Machining processes, dimensional areas,
— Material classes and initial forms,
— Loading time for the manufacturing order,
— Special equipment (tools and attachments),
— Job demands,
— Similarity degree of the parts spectrum.

If a criterion is of special importance for a place of work, e.g. the job demands for the welding of reactor components, then the drawing-up of the programme and distribution of work will be effected according to this criterion. The work programme explained in a detailed manner in the following section is the carrier of the sequence family.

9.3.1 Work Programme for the Distribution of Work

The work programme is the result of a capacity or throughput deadline period. It provides information such as the

— Cost centre which identifies the area of responsibility,
— Planning group which identifies the manufacturing systems,
— Work place number which identifies the individual places of work,
— Part number of the workpieces to be machined,
— Order or batch number of the manufacturing order
— Working operation number,
— Standard time (setting and machining times),
— Final deadline,
— Sequence family number.

The cost centre serves as a limitation of an area of responsibility. It can comprise the machines and places of work belonging to one or more GT-manufacturing systems and is determined by the organizational structure and the cost supervision.

The planning group is used to identify the individual GT-manufacturing systems within the framework of a cost centre.

The place of work number identifies the individual places of work within a planning group.

With this three-stage form of identification all the occuring combinations can be controlled. This form can also be incorporated in the existing system of functional layout and constitutes a smooth transition.

The article number identifies the workpiece to be machined and is based on the single part drawing system. The order or batch number defines the manufacturing order for a specific article number. The working operation number is used to identify the respective working operation in the operational chart.

The standard time includes the setting and machining time. Depending on the time involved, the standard time is stated for 1 or 100 parts. The deadline unit indicates the moment when the workpiece machining has to be completed in the respective planning group. Concerning the GT lines and cells, the deadline is always shown at the final working operation.

The sequence family number is added at the end of the work programme. The sequence family numbers within a planning group (e.g. cell) constitute a code number and provide us with a first impression of the similarity of the manufacturing orders for the respective planning period. Fig. 9.2 shows a work programme for a GT-line, GT-cell and GT-centre with the respective data.

Work Programme for GT-Flow Line				Cost Centre 6830/4		Planning Period 7321						
Part No.	Batch No.	Process No.	Work No.	Standard Time	Deadline	Sequence	Family	Number				
107 070 355 001	015	150	1	29.50		3	013 JH	01	6	2	1	3
		160	2	49.50		3	206 JH	01	8	4	3	3
		170	3	31.00		3	045 JH	01	6	1	1	3
		180	4	31.40	73 216	3	200 JH	01	7	4	3	3

Work Programme for GT-Cell				Cost Centre 7431/2		Planning Period 7321						
Part No.	Batch No.	Process No.	Work No.	Standard Time	Deadline	Sequence	Family	Number				
716 120 219 000	025	010	3	34.00		2	101 KE	20	7	0	5	2
		015	1	69.00		2	205 KE	20	8	1	3	2
		020	4	4.80		2	131 KE	20	3	1	3	2
		025	3	4.80	73 216	2	100 KE	20	3	0	5	2
107 030 790 001	120	005	2	7.50		2	050 BB	10	3	0	5	1
		010	3	1.95	73 216	2	101 BB	10	1	0	5	1
			2	21.50								

Work Programme for GT-Centre				Cost Centre 6342/5		Planning Period 7321						
Part No.	Batch No.	Process No.	Work No.	Standard Time	Deadline	Sequence	Family	Number				
711 200 001 400	220	010	1	17.80	73 213	1	050 NH	03	5	3	7	1
		040	1	27.50	73 226	1	050 NH	03	6	0	7	1
756 125 200 000	010	010	1	6.00	73 216	1	050 JH	02	3	3	7	1
710 123 220 600	045	005	1	13.40	73 214	1	050 IG	02	5	0	7	1
710 129 901 600	003	005	2	53.00	73 215	1	010 GB	01	8	0	7	2
710 121 701 900	001	005	2	80.40	73 216	1	010 AH	20	9	0	5	1

Fig. 9.2. Work Programme for GT-Manufacturing Systems

9.3.2 Information in Respect of Alternative Possibilities

A data sheet, in which the possible sequence family combinations are shown as a coded area for the individual planning groups, will provide information in regard to the technological alternative variant possibilities in the case of deadline bottlenecks. This data sheet can take the form of a card index or can be incorporated in the existing organizational instruments of the production control.

The build-up of the sequence families enables a comparison to be drawn up with the aid of the computer. The purpose of comparison is to find corresponding variants for a certain manufacturing order using the similarity consideration as a basis. In a further step, another comparison has to be made from the loading situation standpoint to enable a decision to be made. Fig. 9.3 shows a coded area to be used as a decision data of a GT-cell. If it is clearly apparent from the work programme of a certain planning period that individual working operations have to be displaced, then a possible alternative variant can be examined by comparing the sequence families with the sequence criteria of the manufacturing systems.

	GT - Cell			
	Coordinate Boring Machine	Horizontal Brass Finisher's Lathe	Openside Vertical Milling Machine	Horizontal Internal Broaching Machine
Main Criteria				
Manufacturing System	2	2	2	2
Machining Process	205	050	101	131
Dimensional Area	I I	0 I	0 I	0 I
Additional Criteria				
Material	0-7	0-7	0-7	0-7
Initial Form	0-9	0-4 7-9	0-9	3
Loading Time per Planning Period	7	8	8	7
Special Equipment	1-4/7/9	1-3/9	1-5/9	1/3/9
Job Demands	3	4	3	2
Similarity Degree of Parts Spectrum	3	1	1	1

Sequence Family Criteria (left vertical label)

Fig. 9.3. Sequence Family Criteria of a GT-Cell

To sum up, it can be said that the studies show that the critical point of GT lies in the work programme preparation and the work distribution. By way of the sequence families a working aid has to be created which will enable the technological similarities of the manufacturing order — in the sense of an optimum sequence — to be considered in a better manner when the work programme is established. A further task is the establishing of alternative variants for the deadline bottlenecks. The sequence families can be incorporated without difficulty in existing production control systems for functional layouts and therefore meet the requirements of both forms of manufacturing systems.

10 Wage Structure

When GT-manufacturing systems are applied in the form of GT-cells and flow lines, the question of individual incentive wage in the case of group work is very much in the foreground. In the case of factories where the GT-manufacturing system has been successfully introduced, this problem factor has been the subject of special discussion, because it has an important influence on the success of the degree of rationalization.

With group work, however, the psychological factors appertaining to equitable payment also have to be considered. In doing so and despite the realization that the group work performance in industrial manufacture is often greater than the sum of the part performances, the individual performance should not be neglected as it still remains and represents the corner stone of every group success. If this influential factor is neglected, it will lead to a leveling of payment and inevitably bring about a reduction in productivity as well.

In a modern wage incentive scheme, the emphasis of the quantity performance and the promotion of the quality performance is especially important, because they both have a decisive influence on the economic working result.

The best known of all the wage incentive schemes is the straight piece-rate scheme. The piece-rate wage is characterized by a directly proportional form of payment with direct influence of the quantity performace. In the case of the incentive wage under piece-rate conditions, an exactly determined progress of working operations is necessary for determining the standard times. As a result of this conditions, piece-rate work is limited in its range of application. The most favourable conditions for piece-rate are to be found in areas where large batch production are involved, where the effort required for the exact pre-planning of the operation sequences and standard times is justified. A periodic check or adaptation to changed factory conditions should be guaranteed, so that the larger time factors, which are caused by minor rationalization measures — particularly in the one-off and small batch production areas — can be limited. In a well-managed firm, these rationalization measures should amount to an average of about 2% per annum. As the quantity performance is only covered by the piece-rate scheme, increased control activities are necassary to maintain the required quality standards.

As a result of increased mechanization and automation, there is therefore a general tendency to reduce the amount of piece work. On the other hand, experience shows that the output in the case of work carried out under day work conditions is less than that attained on an incentive wage basis with measurable reference values. This also applies to day work with performance rating. Comparisons made between firms with and without standard times show output differences of between 10% and 30%.

This situation also shows that the bonus wage system — the third form of incentive wage — has to make special reservations if all including productivity, is to be achieved and maintained in the small and medium batch production and in the special case of GT.

In developing the wage structure, a great deal of importance has been attached to the inclusion of unchanged proven wage components in the extended bonus wage conception, because — apart from the adaptation to the new structure — confidence in a wage scheme must be demonstrated by a retention of a certain constant element. In the development of the bonus wage system the following references have been taken into consideration [45, 53, 54, 55, 56].

10.1 Description of the Bonus Wage System

The bonus wage consists of the standard wage and the bonus share. The standard wage is the individual basic wage which is paid on the basis of an established standard performance, thus the term 'standard wage', and is made up of the following (Fig. 10.1):

— Basic share which is the same for every employee,
— Job evaluation share,
— Behaviour share on the basis of a periodic rating,
— Seniority and age share.

The determination procedure for the standard wage will not be dealt with any further in this work. This build-up depends on the conditions prevailing in the respective country and firm.

The bonus share is based on the average working quantity and quality and is determined with the aid of the bonus wage scheme. The principle is shown in Fig. 10.2 and described in detail in the following.

The left-hand side of this figure represents the quantity influence, whereas the right-hand side shows the quality influence with the resultant bonus factors as performance results of the working quantity and quality.

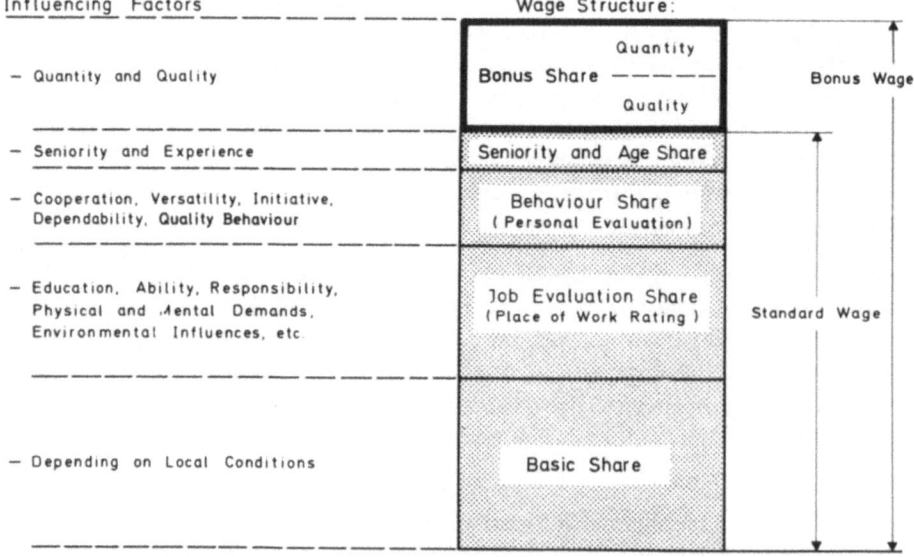

Fig. 10.1. General Structure of the Bonus Wage System

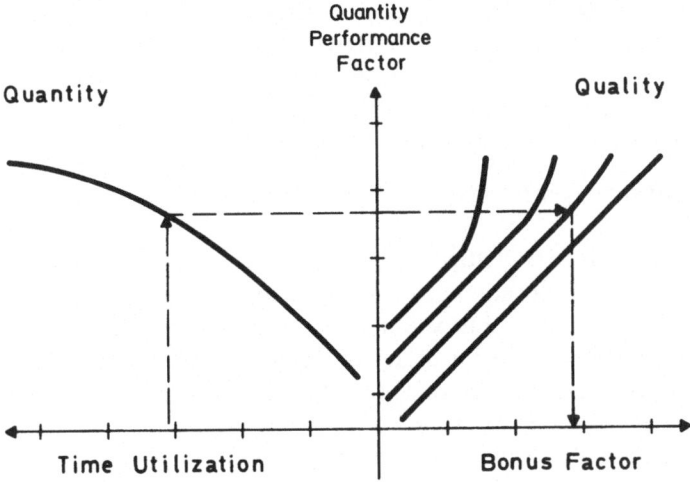

Fig. 10.2. Bonus Wage Scheme

The bonus curves regulate the contrary influence of the two reference values — quantity and quality — and thus enable the performance emphasis to be moved to the respective bonus components which are of economic importance for manufacture.

The economic optimum will change however, according to the way and difficulty of the work.

10.1.1 Quantity Performance

In GT-manufacturing systems, the quantity performance constitutes a significant economic factor for assuring the planned degree of rationalization. As influential factors, the group work in the GT-cell and GT-flow line, the parts structure, the degree of mechanization and automation for the equipment stand thereby in the forefront.

Reference Value

The reference value for the quantity performance is the monthly time utilization factor. In normal cases, it is calculated according to the standard time (or time allowed), the work hours actually performed and the hours actually present. In the case of group work, the personal quantitative performance involvement is given first consideration by using standard time allotments. The calculated time utilization factor (TUF) is thus a measurement of the monthly time utilization, the formula for this being

$$TUF = \frac{\text{standard time (hours)} + \text{day work time per month (hours)}}{\text{time present per month (hours)}}$$

The day work hours per month include, for example, machine cleaning, waiting times etc.

Standard Time Allotment for Group Work

The function of the standard time allotment is to consider the personal performance involvement within the group. This is effected with the aid of the allotment factors for the contributors, which are determined and laid down every month by the supervisors on the basis of a performance rating for those working in the group. The normal work performance is rated with 1.0, a higher performance, for example, the group leader with 1.2. In this way, the personal share of the individual member is correspondingly considered in the light of the group work result. The standard time allocation within the group is obtained by means of the following procedure (Fig. 10.3):

— The standard time of the group per month and the actual time worked constitute the starting basis.
— The individual actual time worked is multiplied by the personal allocation factor (AF) and the result provides the allocation points for the contributors.
— The quotient from the group standard time and the sum of the personal allocation points is the group time factor, which is the degree of the group influence.
— The personal allocation points are then multiplied by the calculated group time factor and this gives us the personal standard time.
— The sum of these personal standard times corresponds again with the group standard time.

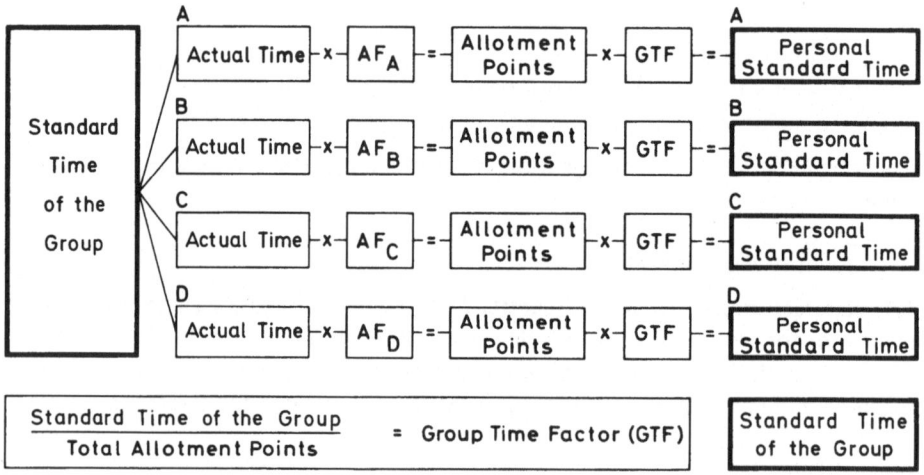

Fig. 10.3. Direct Group Allotment

Determination Procedure for the Quantity Performance

Apart from the personal performance involvement of the employee, the monthly time utilization factor is also dependent on the technical and organizational influence factors of the basic manufacturing system structure, e.g. as is clearly shown when a comparison is made between one-off and large batch production. If the time utilization factor of unchanging employee performance levels is examined over a long period, then we observe statistical distributions with various ranges for the performance results, (Fig. 10.4). This may be attributable to the following technical and organizational factors:

— Manner of the machining task,
— Batch size,
— Degree of difficulty in respect of the work to be performed,

146

- Accuracy of the standard time,
- Degree to which the equipment is mechanized or automated,
- Extent to which the work is planned and the resultant method tolerances for the employee, etc.

In this connection, it should be noted that, in the case of the standard time determination, not only the standard performance had to be taken as a basis, but also the standard technical and organizational factors.

The latter, however, cannot be adapted in every case to the current manufacturing structure. From this, it follows that the time utilization factor cannot be taken as a direct reference value for the quantity performance, for in bonus wage schemes the technical and organizational factors have to be excluded. It is for this reason that the various quantity curves will be introduced for the determination procedure.

The quantity curve allocations are made on the basis of work studies and the statistical evaluation of the time utilization factors. The quantity performance factors resulting from the bonus determination procedure are thus independent of the manufacturing structure and comparable with each other.

The task is now to determine the values for the technical and organizational influences. Manufacturing areas with different manufacturing structures were selected as the basis for investigation. Each manufacturing area employed 40 - 50 men and so the human influence in the areas were more or less governed by the same conditions. Some samples of the evaluations are shown in Fig. 10.5. The resultant time utilization margins are subject to a statistical distribution — the so-called 'Gaussian Normal Distribution'. A normal distribution was established in all the investigated manufacturing areas. The comparison also showed, however, that the average value is not suitable as a reference point for determining the technical and organizational influence values in the determination procedure.

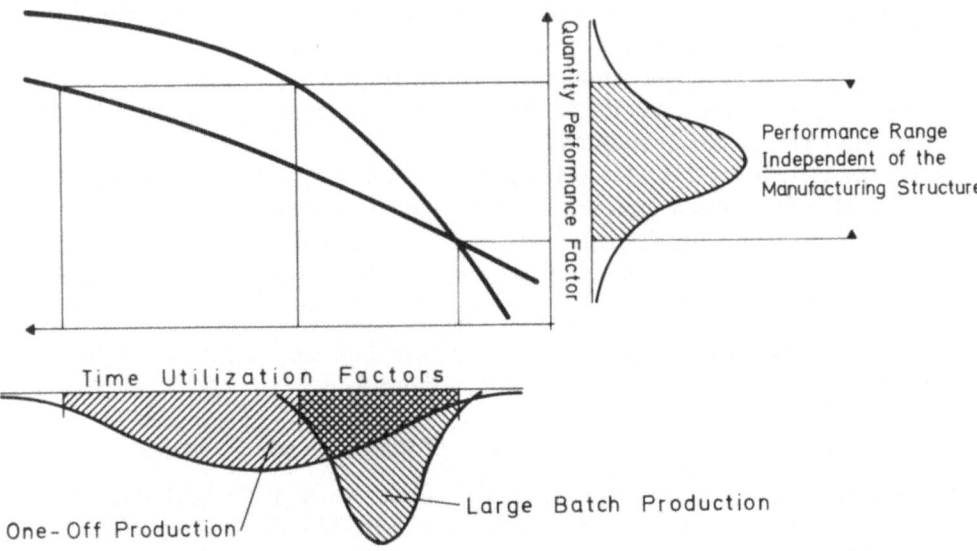

Fig. 10.4. Performance Range Depending on the Manufacturing Structure

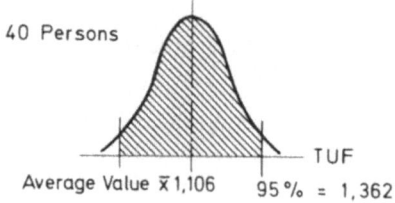

Automatic Lathe Shop

40 Persons

Average Value x̄ 1,106   95 % = 1,362   TUF

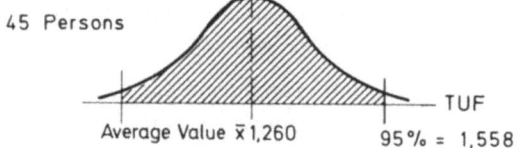

Small Scale Machine Shop (Turning, milling, boring etc.)

45 Persons

Average Value x̄ 1,260   95 % = 1,558   TUF

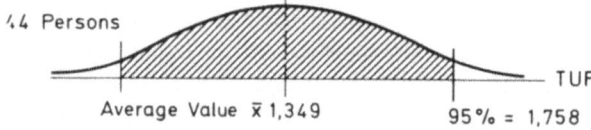

Heavy Machine Shop   (Diesel Engine Assembly)

44 Persons

Average Value x̄ 1,349   95 % = 1,758   TUF

Fig. 10.5. Samples of Distribution of Time Utilization Factor (TUF) of Different Manufacturing Structures

The following points in the bonus wage scheme were established in laying down the bonus curve for the quantity performances (Fig. 10.6):

— The quantity grades are determined by the upper level of the time utilization margins (95%) for the individual manufacturing areas and checked every six month as a result of the statistical evaluation of the time utilization factors.
— For the purpose of simplification, nine quantity stages with corresponding classification ranges are laid down in the model. These have also proved to be successful during tests under practical conditions.
— The bonus curve for the individual stages is determined by the upper and lower limit values of the admissible human and economic performance margins.
— In the lower area, the function is progressive for the purpose of emphasising the performance, whereas in the upper area the aim is to increase the stress on the quality side.

Thus the quantity performance factor resulting from the bonus determination procedure reflects the personal influenced quantity performance.

Quantity Performance within the Framework of GT-Manufacturing Systems

In the determination procedure for the quantity performance, the individual GT-manufacturing system constitutes a self-contained bonus area. With the GT-manufacturing cells and flow lines, the quantity performance is calculated on the basis of the group standard time. The GT-centre which is operated by one man in normal circumstances,

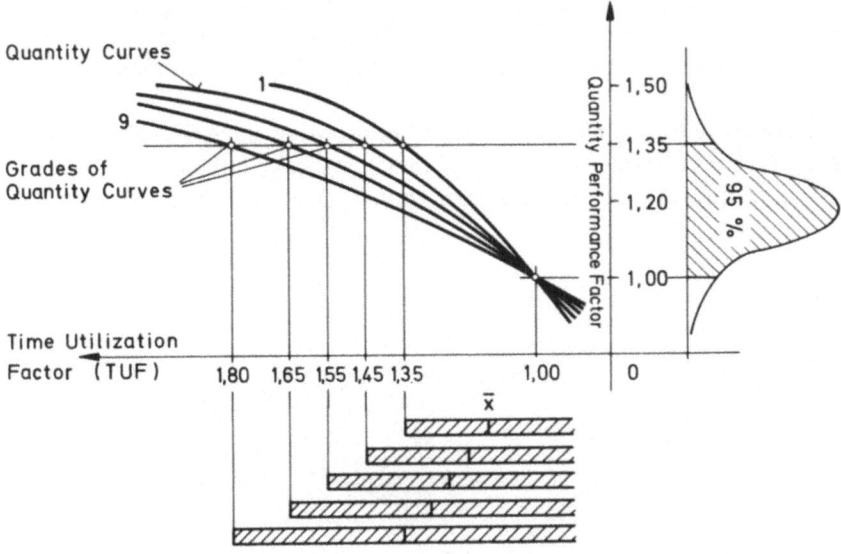

Fig. 10.6. Model of Quantity Performance Determination

is the subject of special considerations, except in the case of shift work or larger objects, where two or more people are required.

In the case of one-man operation, the determination procedure is applied on the basis of individual standard times without using the allocation formula for group work.

The following pictures briefly explain the quantity curves for three typical basic forms:

— A steep quantity curve for e.g. a GT-flow line and with high degree of similarity of the parts spectrum because the performance range, caused by the objective influences in the case of a parts spectrum with a high degree of similarity, is small (Fig. 10.7).

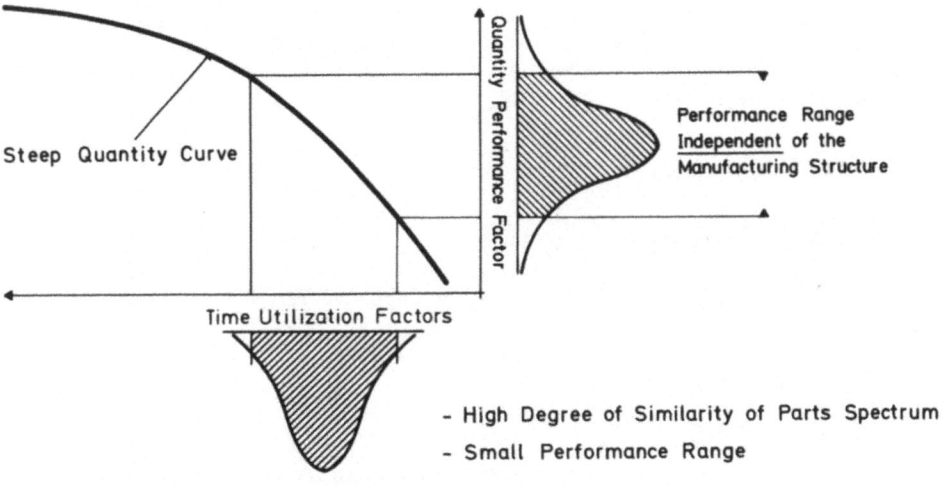

Fig. 10.7. Quantity Curve of a GT-Flow Line

- A flattened out quantity curve for e.g. a GT-cell and with medium degree of similarity of the parts spectrum, because the performance range, caused by the objective influences in the case of a parts spectrum with a medium degree of similarity, is larger (Fig. 10.8).
- A flat quantity curve for e.g. a GT-centre with low degree of similarity of parts spectrum, because the performance range, caused by the objective influences in the case of a parts spectrum with a low degree of similarity, is large (Fig. 10.9).

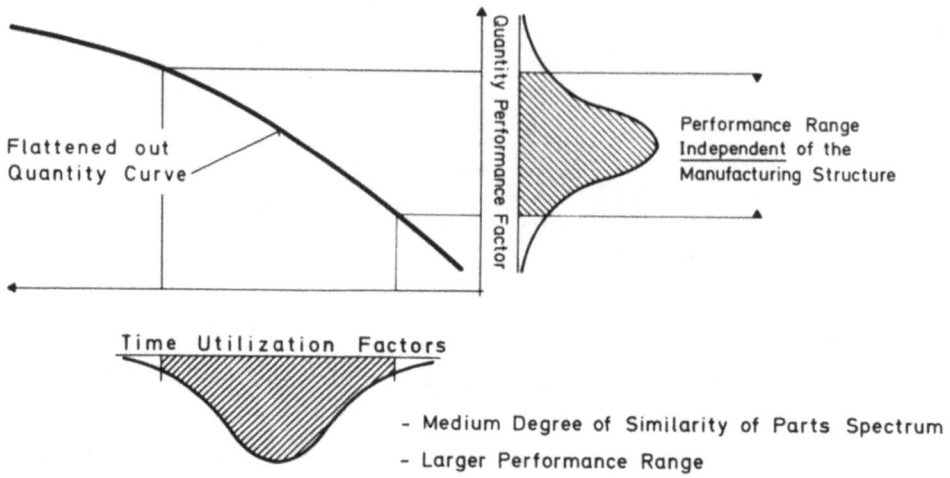

Fig. 10.8. Quantity Curve of a GT-Cell

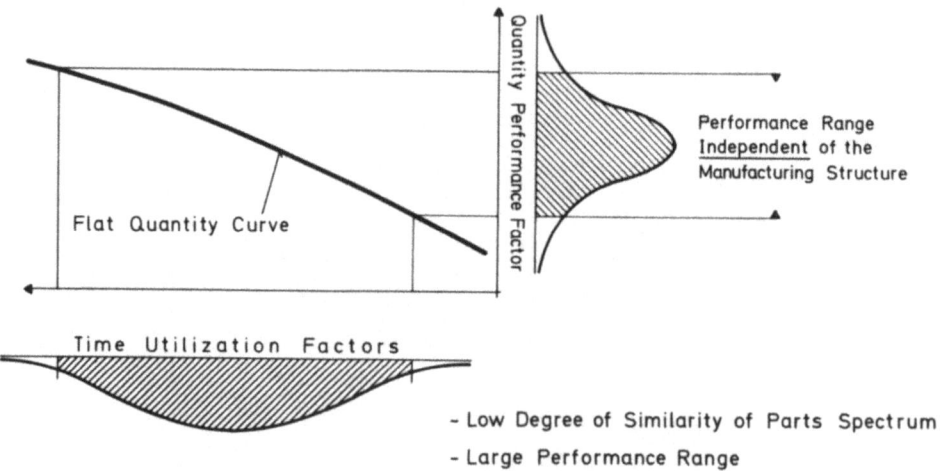

Fig. 10.9. Quantity Curve of a GT-Centre

The performance range will become smaller and the quantity curve steeper with increase of the degree of similarity in the parts spectrum and the degree of automation.

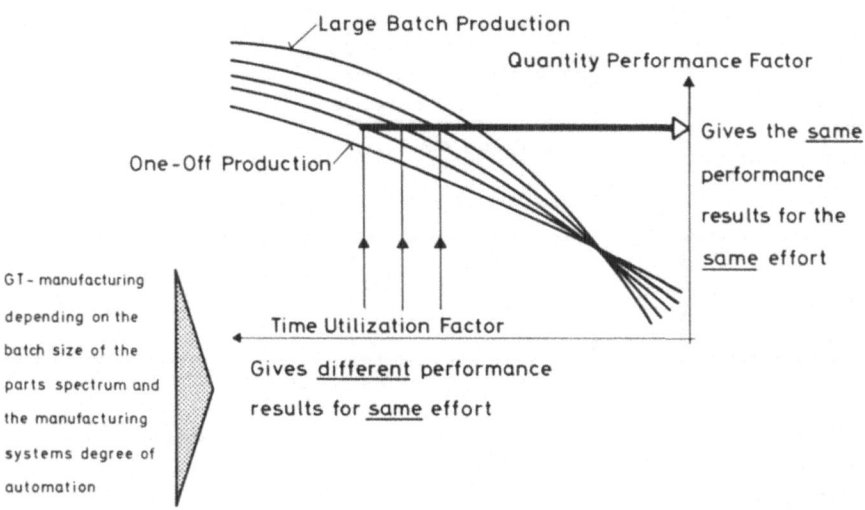

Fig. 10.10. Determination Procedure for the Quantity Performance

In Fig. 10.10 the framework of the determined quantity curves for the individual GT-manufacturing systems is summarized as well as the resultant comparative quantity performance, which is still only dependent on the personal performance involvement.

10.1.2 Quality Performance

From the wage standpoint, the quality of work is influenced by the following factors which have to be considered in any modern bonus wage concept:

— Quality demand or requirement,
— Quality behaviour,
— Quality performance.

The quality demand includes the ability, aptitude and experience needed to produce the required quality on a certain place or work. These characteristics are considered by means of the framework of the job evaluation and influence the extent of the job evaluation share as far as the standard wage is concerned.

A further influential factor is the quality behaviour. It poses the question 'how', i.e. in which manner does the employee perform his work. Does he show a readiness to produce quality? Is his performance curve constant or does it fluctuate? Further important factors are the efforts made towards inspection and instruction and the ability of the employee for self-inspection duties. These characteristics are checked periodically by means of a behaviour rating through a committee consisting of the works manager, foreman, inspector and a personel department representative.

The quality performance includes the specific quality characteristics such as accuracy to size and places the word 'what' in the foreground i.e. what quality result was attained by the employee during a certain time period, e.g. say a month. In Fig. 10.11 the three influential factors of the quality are shown as well as their integration into the overall wage system.

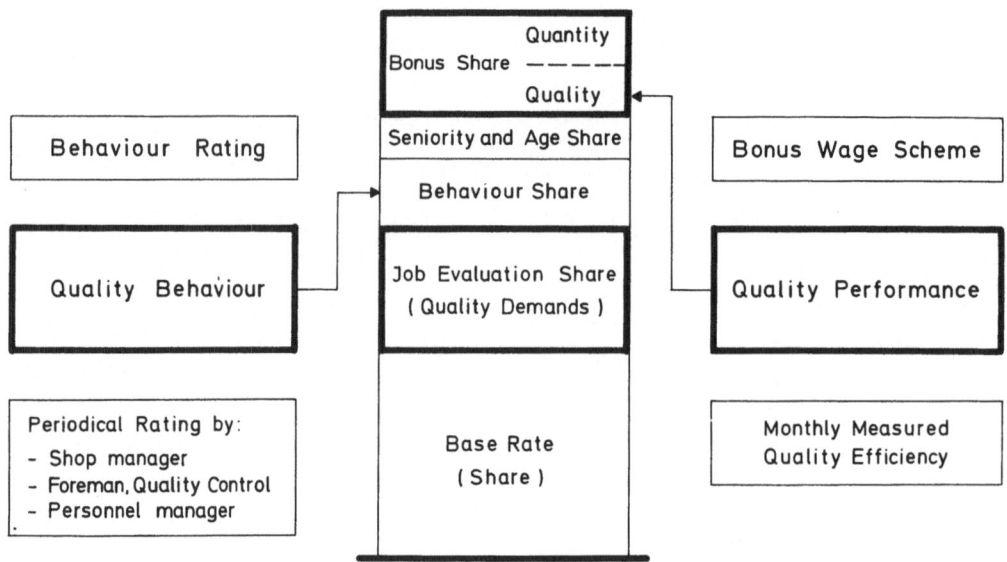

Fig. 10.11. Influential Factors of Quality

Determination Procedure for the Quality Performance

The quality performance is determined with the aid of quality curves. From the economic standpoint, the quality curves are laid down in relation to the quality demands of the individual places of work. With the quality demands, the difficulty and the degree of influence on the quality performance and indirectly the economic components of performance are considered. For the purpose of simplification, the following areas of demand have been established: simple, medium, special and highest quality demands (Fig. 10.12). The parameter for determining the quality performance are:

— The quantity performance factor resulting from the determination procedure for the quantity performance.
— The quality curve of the quality performance personally achieved by the employee, depending on the allocated quality area of demand.

In the case of high quality work, the emphasis of performance can thus be increasingly directed towards the quality side, whereas with work where simple quality demands are involved, the assurance of quality stands beside the work quantity in the foreground.

Quality Characteristics

The choice of quality performance characteristics should not be limited to dimensional and surfaces tolerances only, but should also include the economic and cost factors that may be influenced by the quality performance. For example, this includes the number of rejects, the defect reports, inspection costs in comparison with manufacturing and other quality characteristics that are needed nowadays for sucessful management. Fig. 10.13 shows a number of quality characteristics concerning the accuracy and technological execution of the working operations. It permits a selection for the individual places of work.

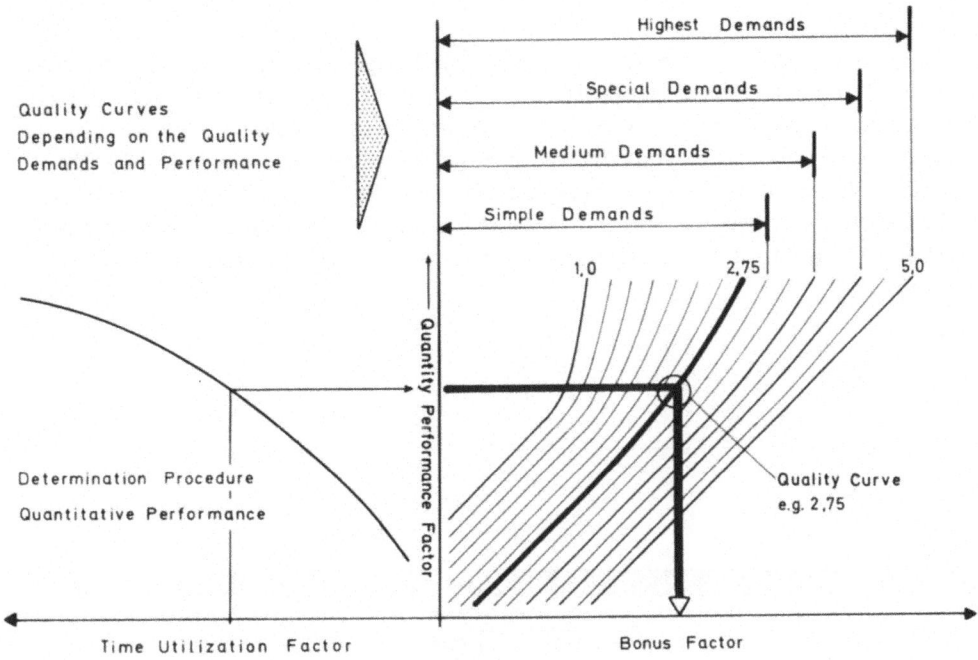

Fig. 10.12. Determination Procedure for the Quality Performance

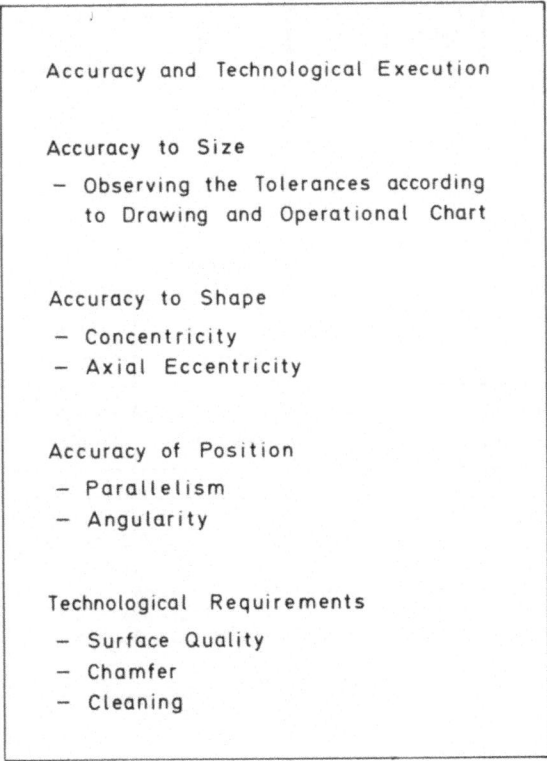

Fig. 10.13. Quality Criteria (Example)

153

Quality Determination

The quality is determined by the management and inspection staff positions. The quality determination is based on individual random checks, which can consist of a complete batch, the working operation on a workpiece or, in the case of extensive working operations, a partial working operation. The test investigations showed that 20 - 30 random checks should be made every month if a conclusive basis is to be obtained. In the case of self-inspection, only individual random checks were made. These were principally confined to the final inspection where a check was made on the quality performance.

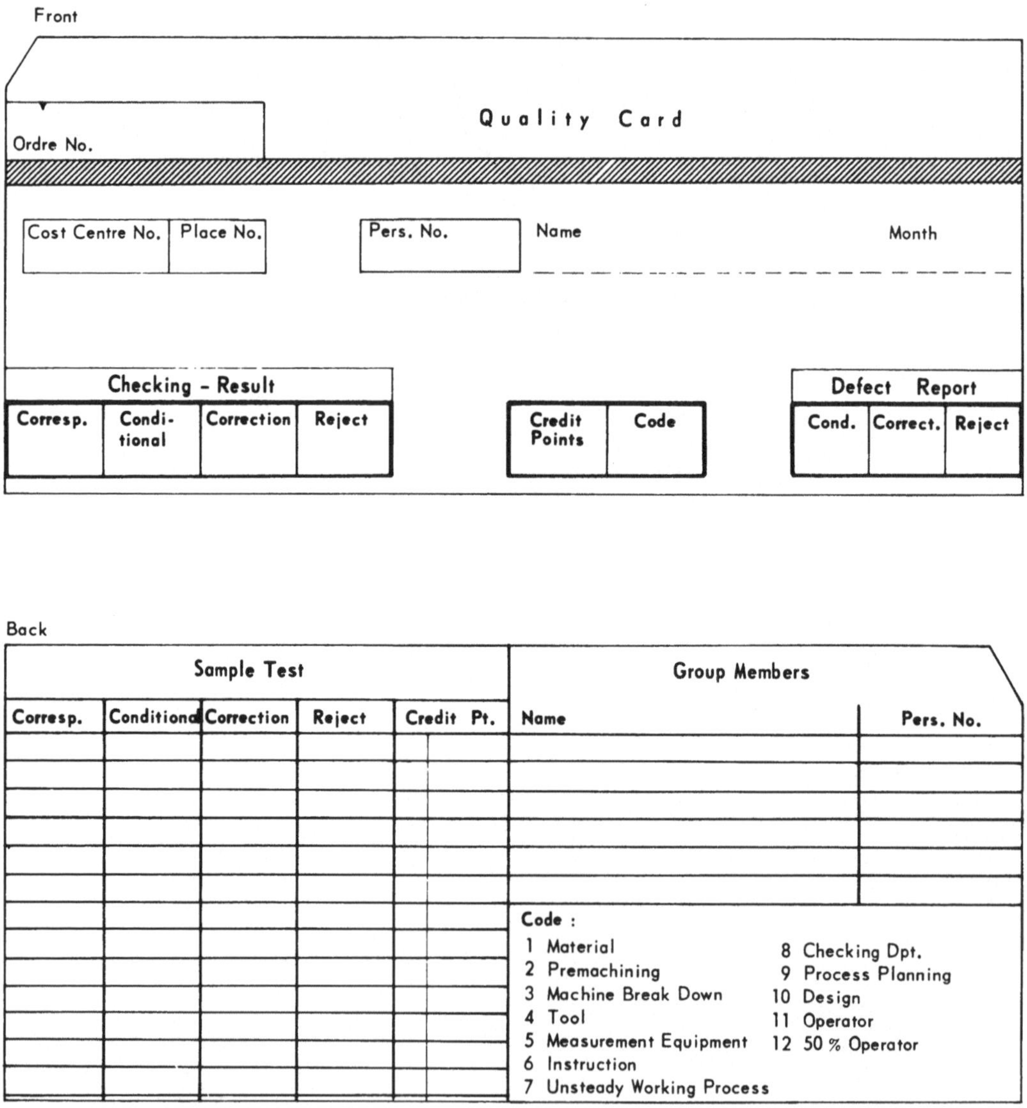

Front

				Quality Card		
Ordre No.						

Cost Centre No.	Place No.		Pers. No.	Name		Month

Checking – Result

Corresp.	Condi-tional	Correction	Reject		Credit Points	Code		Cond.	Correct.	Reject

Defect Report

Back

Sample Test					Group Members	
Corresp.	Conditional	Correction	Reject	Credit Pt.	Name	Pers. No.

Code :

1 Material	8 Checking Dpt.
2 Premachining	9 Process Planning
3 Machine Break Down	10 Design
4 Tool	11 Operator
5 Measurement Equipment	12 50 % Operator
6 Instruction	
7 Unsteady Working Process	

Fig. 10.14. Quality Recording

The results of the random checks are classified according to the following quality classes:

- *Corresponding:* The random check result corresponds with the quality established on the drawing and in the work instructions.
- *Conditional:* The result only conforms under certain conditions, the test values lie on the limits of accuracy to size and execution.
- *Correction:* The result calls for correction or re-work of a modification of the parts.
- *Reject:* The random check result shows that the test piece is not suitable or can only be used with extensive effort.

The classification of the random checks into quality classes is made independently of the personal and objective influence factors. If a negative quality result is attributable to objective reasons, the responsible line manager may make a subsequent correction in the form of plus points — which are explained in the section quality evaluation. In this way, only the personal influence on the quality performance is considered. The cause code for the objective influence factors is shown on the quality card used for recording the random check results (Fig. 10.14).

A collective list (Fig. 10.15) showing all the random checks may be used instead of the quality card. The form to select should be based on the existing inspection procedure. In doing so, it should be worked with an inspection plan, because the quality examination is not comprehensive but should be based on an objective-orientated random check so as to reduce the inspection effort in areas with high quality levels.

Month	Quality Classification				
Random Check	Corre-sponding	Condi-tional	Correc-tion	Reject	Credit Points
1	I				
2	I				
3			I		I
4	I				
5	I				
6		I			
7	I				
8	I				
9			I		
10	I				
11		I			
12		I			
13	I				
14				I	I
27	I				
28					I
29		I			
30	I				
31			I		
Total	23	4	3	1	3

Fig. 10.15. Quality Determination in the GT-Cell

155

Quality Evaluation

The monthly random check record constitutes the basis for the quality evaluation. Weight factors are provided to obtain a stronger differentiation between the individual quality classes. These factors and the evaluation process are shown in Fig. 10.16. In laying down the weighting factors, it was assumed that the main weighting will be on the quality class corresponding and the resultant intermediate value is expressed in the form of quality points. The solution was selected with a view to strengthening the quality objective. Schemes, which are only based on the recording of defects and do not determine the positive quality results, often lead to the same end result. Admittedly, they require a smaller amount of administration effort, but they have the great psychological disadvantage of only referring to defects and therefore influence the work motivation in a negative manner.

Quality Classification	Number	Weight Factor	Quality Points	Credit Points
Corresponding	23	4	92	
Conditional	4	3	12	1
Correction	3	1	3	1
Reject	1	0	0	1
Total	31		107	3

$$\text{Quality Value} = \frac{\text{Quality Points} + \text{Credit Points}}{\text{Number of Random Checks per Month}} = \frac{107 + 3}{31} = 3{,}45$$

Fig. 10.16. Quality Evaluation

The crediting of points for the negative random check results caused by objective influence factors such as material defects, etc., is recorded under the column 'credit points' (Fig. 10.16). The quotient from the sum of the quality points and the credit points over the number of random checks per month produces the quality value.

As opposed to a summarized form of quality evaluation — as is used in other schemes — the quality value represents the total individual result and, from a statistical standpoint, is a more conclusive and more objective basis for the bonus determination.

The quality value, however, does not yet consider the quality demands. The advantage of this is that in a GT-cell, for example, the quality value may be determined for the complete working group and the differentiation, e.g. between the group leader and the auxiliary workers, can be made with the aid of quality demand areas, as is shown in the next section.

Quality Grading

The function of the quality grading is to allocate monthly quality values in relation to the four defined quality demand areas (simple, medium, special and highest demands).

To simplify matters, the quality value of the individual demand is classified into value areas and allocated to a quality grade. The value areas for work with simple or highest quality demands are shown in Fig. 10.17.

The monthly quality grade evaluated in this way reflects the quality performance attained and provides the reference values for determining the combined bonus factor for the work quality and work quantity.

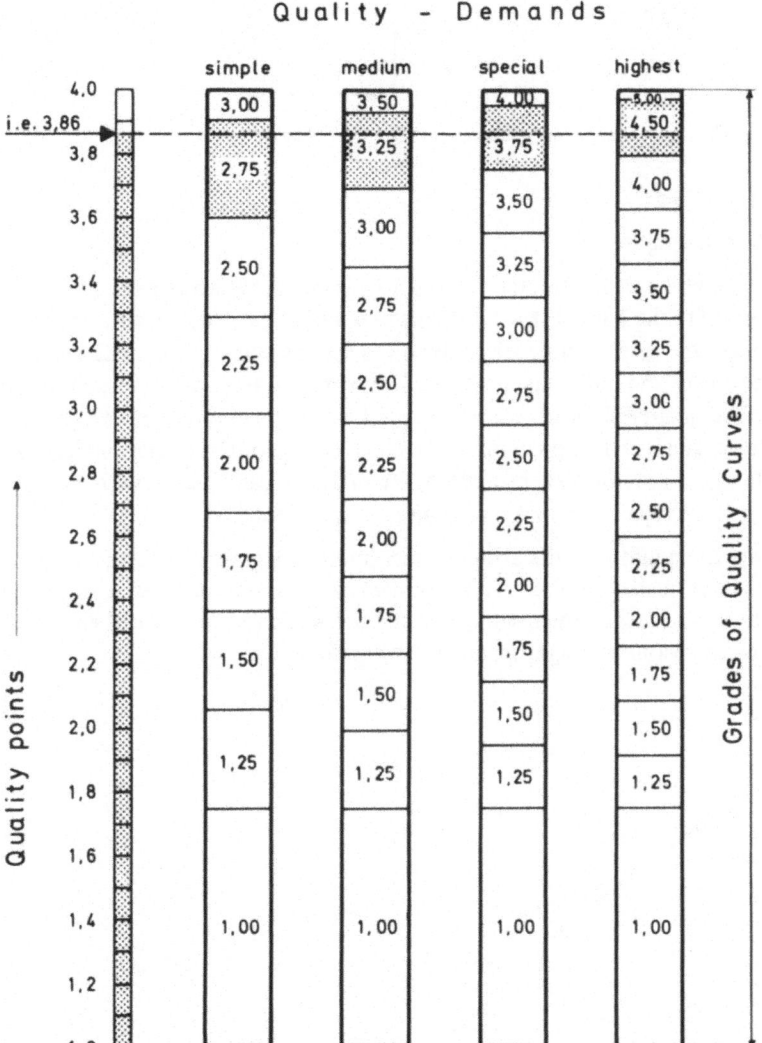

Fig. 10.17. Quality Grading in Respect to the Demands

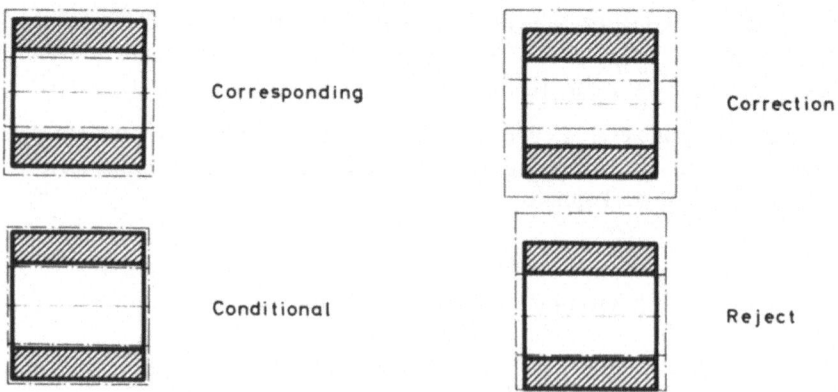

Fig. 10.18. Forging Allowance and Quality Classification

10.1.3 Application Example

The following example is taken from the forging area. Fig. 10.18 shows a section of the representative parts. On the basis of the work analysis and in accordance with the various material groups, the optimum forging tolerances for the subsequent cutting operations were laid down for the individual parts, and allocated to the quality class allotted 'corresponding'. The quality class, 'correction' is used when the component is oversized, which results in a very large cutting volume. The class 'conditional' is determined when the optimum tolerances are not met, however the workpiece can be used with special care. With quality class 'reject' the minimum tolerances are not met.

Fig. 10.19 presents the quality results after the test phase in a graphical form and also shows the influence of quality on the bonus payment. If it is considered that an average reduction of 25% in the costs of the subsequent cutting work was realized, then this example will show the economic significance of the quality side.

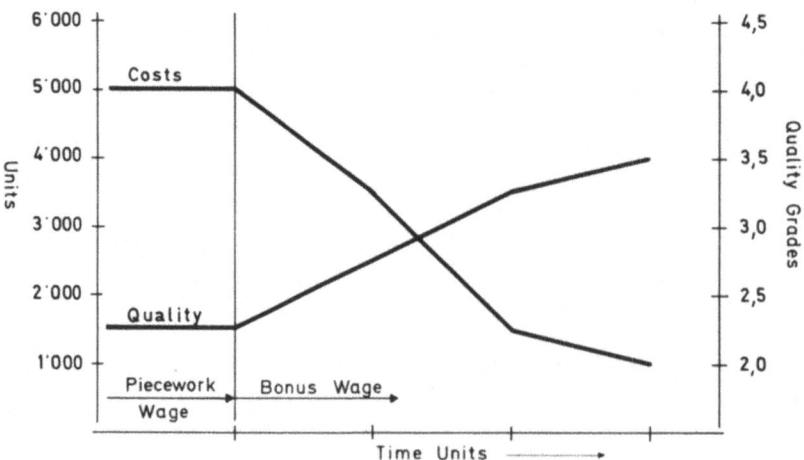

Fig. 10.19. Costs and Quality Performance after Introducing the Bonus Wage System

158

10.1.4 Handling of the Bonus Wage Scheme

A differentiation must be made between the system fundamentals, the system concept and the system handling. A car may be used as an example. The majority of drivers know how to operate the vehicle and are also familiar with its driving characteristics and performance, on the other hand, very few experts know about the design and construction.

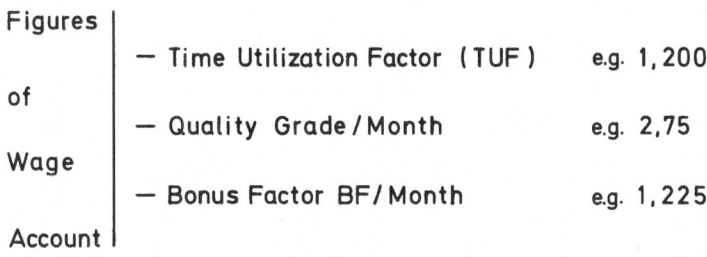

```
Figures
         |  — Time Utilization Factor  (TUF)      e.g. 1,200
of       |
         |  — Quality  Grade / Month             e.g. 2,75
Wage     |
         |  — Bonus Factor BF / Month            e.g. 1,225
Account  |
```

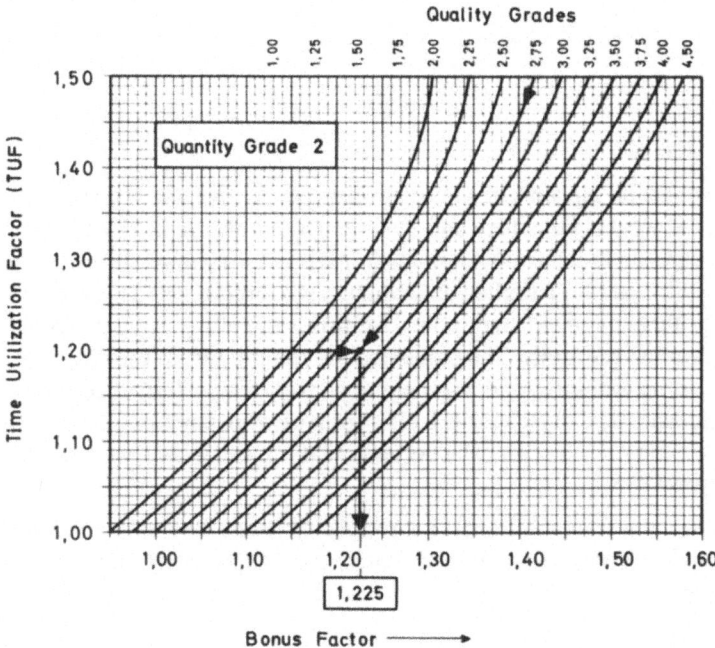

Fig. 10.20. Diagram for Examination of the Monthly Bonus Factor BF

In our case, the employee should be provided with simple instruction, which describes the specific performance factors of the respective place of work, and a tailor-made determination diagram for working out the monthly bonus factor. Such a diagram is shown in Fig. 10.20. It represents a section of the whole system and refers to a certain case of application. On the basis of an established quantity curve and two reference values — time utilization factor and quality grade — the monthly bonus factor can be determined with the bonus account without any difficulty (Fig. 10.21).

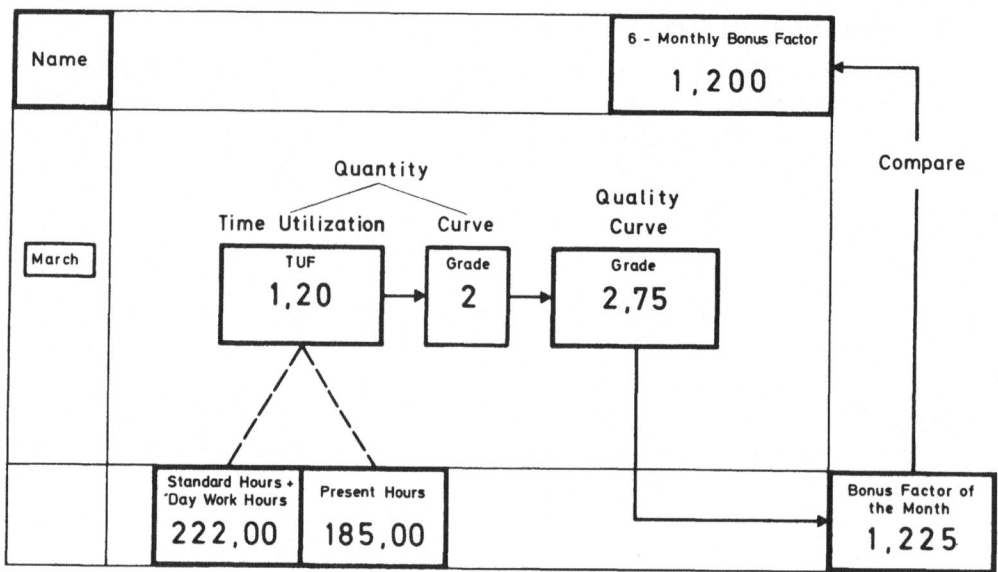

Fig. 10.21. Comparison for the Bonus Factors in the Wage Account

Conclusion

This bonus wage scheme fulfils the aims of the performance incentive and assures the performance in respect of working quantity and quality. The advantage of this bonus wage scheme is that it can be atuned to the individual manufacturing system forms and places of work, but is still based on a unified and comparable foundation. This is very important from the equitable payment standpoint, for this is not always guaranteed in the case of specific individual solutions. The bonus factors resulting from the diagram enable the bonus wage to be evaluated and also help to control the performance and assure the planned rationalization object for GT. The form of wage payment should always be adapted to the local conditions. The system build-up is designed for manual and computer aided operations.

Apart from the systems engineering solution, other social conditions have to be considered. These may vary from country to country and represent an object of negotiation among the social partners.

11 Conclusions

The application of systems engineering has been shown to be a rational and comprehensive method of work for the integrated planning and development of a GT-concept. It enables improvements to be made in the overall efficiency of GT in the field of mechanical engineering with complex production programmes.

The proposed method for implementation of GT in a machine building company consists of a general level, which contains the overall aspects, and a specific level that duly considers the special partial aspects of GT. This approach has proved to be expedient in the development of procedures and working aids, and also for practical application.

The principle of a three-stage method for the standardization and rationalization of planning, design and manufacture, on the basis of a similarity and frequency analysis gives the concept a flexible character which enables a stepwise build-up of GT in mechanical engineering companies.

The proposed technological classification system based on three elements, namely

— Workpiece,
— Operation,
— Equipment,

constitutes a general and comprehensive instrument for analysing the different crucial points of rationalization of GT in planning, design and manufacture.

The collection of data on the basis of representative product types enables a considerable reduction in the coding effort.

Investment and layout planning have to consider two reasons for the deviation of the effective parts spectrum from the planned one. The first reason is the annual modification of the product quantity, the second one the product mixture. Both factors can be determined with the aid of so-called representative product types, on the basis of trend studies and variant calculations.

The build-up of a 'technological data bank' with stored classification and identification data for the representative product types will enable information to be obtained much quicker than by conventional data collecting methods.

Application-orientated evaluation programmes enable various alternatives to be simulated on the same computer run. This shortens the computer runs and improves the accuracy of the planning data for the individual rationalization tasks.

In an iterative process, the described procedure for layout planning enables the optimum stage of the three defined GT-manufacturing system:

— GT-centre,
— GT-cell,
— GT-flow line,

to be determined with the aid of similarity and loading analysis.

With regard to parts design the integrated approach of standardization and rationalization enables a larger potential number of components to be covered. Complimentary to the standard and recurring parts, which form a comparatively small percentage of the total, it is proposed that similarity types be formed which would enable a larger proportion of total parts to be considered. These similarity types are sub-divided according to the degree of similarity and frequency of occurence into

— Main types,
— Basic types,
— Single types.

The procedure is described with examples taken from practice.

By using the three-stage method for the build-up of the similarity data for process planning and work measurement, a flexible instrument is created which will consider the economic aspects of both effort and return.

As a further element in this concept, the sequence families aid the production control in the optimization of the overall throughput and in establishing alternative manufacturing systems in bottleneck situations.

The final link in the concept is the wage structure. With the aid of the developed bonus wage system with the elements

— Work quantity and quality,
— Allotment factors for group work,
— Determination of the bonus curves in relation to the similarity characteristics of GT-manufacturing systems,

it is possible to utilize in an optimum manner the economic advantages of GT in manufacturing and, at the same time, obtain an equitable payment.

To sum up, the concept represents a comprehensive and flexible approach for the introduction and integration of GT into a company's production system.

References

1. Mitrofanow, S.P.: Wissenschaftliche Grundlagen der Gruppentechnologie. VEB Verlag Technik Berlin 1960
2. Thornley, R.H.: Group Technology – A complete manufacturing system. Inaugural Lecture 14th Oct. 1971. University of Aston, Birmingham
3. Opitz, H.: Produktplanung – Konstruktion – Arbeitsvorbereitung. W. Giradet, Essen 1971
4. Burbidge, J.L.: Group Technology. Proceedings of International Seminar September 1969. International Centre for Advanced Technical and Vocational Training, 140, Corso Unita d'Italia, Turin
5. Koenigsberger, F.: The Use of Group Technology in the Industries of Various Countries. Annals of the CIRP, Vol. 21/2, 1972
6. Edwards, G.A.B.: Readings in Group Technology. Machinery Books, 1971
7. Thornley, R.H.; Middle, G.H.; Connolly, R.: Gruppentechnologie – ein Weg zur Lösung von Fertigungsproblemen in der losgebundenen Fertigung. Industrie-Anzeiger No. 42 and 51, 1971
8. Edwards, G.A.B.; Koenigsberger, F.: Group Technology – The Cell System and Machine Tools. The Production Engineer July/August 1973
9. Opitz, H.; Wiendahl, H.P.; Grabowski, H.: Investitionsplanung einer Maschinenfabrik unter Berücksichtigung der Werkstückanforderungen. wt – Z. ind. Fertig. Vol. 61 (1971), No. 1
10. Klügl, P.: Eine Chance für alle im Folkshemmet – 'Gerechtere' Bildung und menschlichere Arbeitsplätze. Die Weltwoche, Zürich, 31st Jan. 1973, No. 5
11. van Beek, H.G.: Neue Formen der Arbeitsorganisation bei Philips. VSBI-Jubiläumstagung 4th Oct. 1973, ETH Zürich
12. Lattmann, Ch.: Das norwegische Modell der selbstgesteuerten Arbeitsgruppe. Paul Haupt, Bern
13. Collective List of Authors: Systems Engineering. Seminar am Betriebswissenschaftlichen Institut Zürich, Nov. 1972
14. Brankamp, K.; Wiendahl, H.P.: Rationalisierungsmöglichkeiten im Konstruktionsbereich. VDI-Bericht No. 152, Düsseldorf 1970
15. Kesselring, F.; Arn, E.: Methodisches Planen, Entwickeln und Gestalten technischer Produkte. Konstruktion Vol. 23 (1971), No. 4
16. Dähnert, H.: Die vier Stufen der Programmiertechnik. Fertigung 1970, No. 2/3
17. Budil, W.: Planung, Koordination und Kontrolle als betrieblicher Ablauf. Industrielle Organisation. Vol. 38 (1969)
18. Büchel, A.: Systems Engineering. Industrielle Organisation, Vol. 38 (1969)
19. Zangemeister, Ch.: Systemtechnik – eine Methodik zur zweckmäßigen Gestaltung komplexer Systeme. Zeitschrift für Organisation, Vol. 5 (1970)
20. Chestnut, H.: System Engineering Tools. John Wiley & Sons, New York 1965
21. Becker, A.M.: Operations Research: Ein Hilfsmittel zum Planen, Rationalisieren und Optimieren im Betrieb. Technische Rundschau Sulzer, Vol. 52 (1970) No. 2
22. Gombinski, J.: The Brisch Classification and Group Technology in 'Group Technology' by J.L. Burbidge. 1969. International Centre for Advanced Technical and Vocational Training. 140, Corso Unita d'Italia, Turin
23. Opitz, H.: Werkstückbeschreibendes Klassifizierungssystem. W. Giradet, Essen 1968
24. Arn, E.; Dähnert, H.: Technologisches Strukturplanungssystem. Fertigung, Vol. 1970, No. 4
25. Hahn, R.; Kunerth, W.; Roschmann, K.: Die Nummerung im Fertigungsbetrieb. Werkstatt-technik Vol. 58 (1968) No. 11
26. Zimmermann, D.: ZAFO, eine allgemeine Formenordnung für Werkstücke-Gestaltung, Handhabung und Rationalisierungserfolg. G. Grossmann, Stuttgart 1967
27. Collective List of Authors: Group Technology and Numerical Control in Japan. 1969 Group Technology-Numerical-Control Study Team. The Japan Institute of Industrial Engineering. J.P.C. Building 1 - 1, 3-Chome, Shibuya-Ku, Tokyo

28. Kindhauser, G.E.: Klassifikation, Werkzeugmaschinen und Werkzeuge. Fachmesse für Werkzeugmaschinen und Werkzeuge, FAWEM 68, Basel 1968

29. Brankamp, K.; Olbertz, H.; Schütze, R.: Aufbau und Anwendungsmöglichkeiten eines Klassifizierungssystems für Schweißeinzelteile. Schweißen und Schneiden, Vol. 22 (1970), No. 12

30. Rockstroh, W.: Technologische Betriebsprojektierung Gesamtbetrieb. VEB Verlag Technik, Berlin 1971

31. Wiendahl, H.-P.: Technische Struktur- und Investitionsplanung. W. Giradet, Essen 1973

32. Aggteleky, B.: Fabrikplanung. Optimale Projektierung, Planung und Ausführung von Industrieanlagen. Carl Hanser, München 1971

33. Wagner, R.: Arbeitsgangverschlüsselung. G. Grossmann, Stuttgart 1971

34. Eversheim, W.; Wiendahl, H.P.: Rationelle Auftragsabwicklung im Konstruktionsbereich. W. Giradet, Essen 1971

35. Simon, R.: Rechnerunterstütztes Konstruieren. Dissertation TH Aachen, 1968

36. Clausen, U.: Konstruieren mit Rechnern. Springer, Berlin 1971

37. Steinmetz, G.: Integrierte Konstruktion und Arbeitsplanung für Varianten. Dissertation TH Aachen, 1973

38. Opitz, H.: Rationalisieren in der Arbeitsplanung. 12th/13th June 1973. Technische Akademie, Wuppertal

39. Olbrich, W.: Arbeitsplanerstellung unter Einsatz elektronischer Datenverarbeitungsanlagen. Dissertation TH Aachen, 1970

40. Durie, F.R.E.: A Survey of Group Technology and its Potential for User Application in the UK. The Production Engineer, Febr. 1970

41. Knauber, K.: Bildung von Bearbeitungsfamilien mit Formen- und Bearbeitungskennziffern, Teile 1 und 2. TZ für prakt. Metallbearbeitung Vol. 66 (1972) No. 12 and Vol. 67 (1973), No. 1

42. Tuffentsammer, K.: Form- oder fertigungsorientierte Teileordnung? Maschinenmarkt, Vol. 73 (1967), No. 96

43. Collective List of Authors: Introduction for Workstudy. 2. Ed. International Labor Office, Geneva (1971)

44. REFA: Methodenlehre des Arbeitsstudiums, Teil 2: Datenermittlung. Carl Hanser, München 1972

45. Riggs, J.L.: Production Systems: Planning, Analysis and Control. John Wiley & Sons, New York

46. Wojda, F.A.: Modell der Arbeitsvorbereitung unter Berücksichtigung der Einzel- und Serienfertigung, Habilitationsschrift, TH Wien 1972

47. Schweizerische MTM-Vereinigung: MTM-Information No. 1 - 8. MTM-Sekretariat, Rue des deux-marches, Vevey

48. Hediger, P.: Einsatz des Computers und vorbestimmter Zeitelemente als Rationalisierungswerkzeuge zur Fertigungsplanung. Information Meeting, 20th/21st March 1973, Zürich

49. Koenigsberger, F.; Caudwell, F.W.; Haworth, E.A.; Levy, H.H.: Verbesserter Produktionswirkungsgrad durch eine computerunterstützte Produktionssteuerung im Zusammenhang mit Teilefamilienfertigung. Fertigung No. 5/1972

50. Dill, P.: Produktionsplanung und Teilefamilienfertigung. Industrie-Anzeiger, Vol. 87 (1965) No. 103

51. Schwander, A.: Automatische Produktionssteuerung. Planung und Produktion No. 12/1972

52. Meyer, B.E.: Beitrag zur systematischen Auswahl von Betriebsmitteln auf der Grundlage von Gruppengesetzmäßigkeiten. Dissertation ETH Zürich, 1973

53. Soom, E. Grundlage einer universellen Leistungsentlohnung für Stunden- und Monatslöhner. Europäischer Verband für Arbeitsstudien, Jahreskongress 2nd - 4th June 64, Zürich

54. Euler, H.; Stevens, H.: Vorschlag für eine neue Methode der Leistungsentlohnung. Stahl und Eisen, Düsseldorf 1962

55. Baierl, F.: Produktivitätssteigerung durch Lohnanreizsysteme. Carl Hanser, München 1965

56. Arn, E.: Leistungslohn- und Auszahlungsarten für das Personal im Stunden- und Monatslohn. Schweizerische Arbeitgeber-Zeitung Vol. 63 (1968) No. 6